A REVISION OF THE MARINE NEMATODES OF THE SUPERFAMILY DRACONEMATOIDEA FILIPJEV, 1918 (NEMATODA: DRACONEMATINA)

A REVISION OF THE MARINE NEMATODES OF THE SUPERFAMILY DRACONEMATOIDEA FILIPJEV, 1918 (NEMATODA: DRACONEMATINA)

BY

M. W. ALLEN and E. M. NOFFSINGER

UNIVERSITY OF CALIFORNIA PRESS
BERKELEY · LOS ANGELES · LONDON

UNIVERSITY OF CALIFORNIA PUBLICATIONS IN ZOOLOGY

Volume 109
Approved for publication June 24, 1977
Issued September 15, 1978

UNIVERSITY OF CALIFORNIA PRESS
BERKELEY AND LOS ANGELES
CALIFORNIA

UNIVERSITY OF CALIFORNIA PRESS, LTD.
LONDON, ENGLAND

ISBN: 0-520-09583-9
LIBRARY OF CONGRESS CATALOG CARD NUMBER: 77-83108

© 1978 BY THE REGENTS OF THE UNIVERSITY OF CALIFORNIA
PRINTED IN THE UNITED STATES OF AMERICA

CONTENTS

Foreword, by A. R. Maggenti ... viii
Acknowledgments ... 1
Introduction ... 1
Biology ... 2
    Habitat Associations (Table 1) ... 2
    Incidence of Suctoria (Table 2) ... 3
Known and Recorded Distribution of Draconematoidea ... 5
    World Map ... 5
Abbreviations Used in Species Descriptions and Keys ... 6
Explanation of Measurements Used in Species Descriptions and Keys ... 7
Morphology ... 9
    Cuticle ... 9
    Setae ... 10
    Lips ... 10
    Cephalic Sensory Organs ... 10
    Amphids ... 11
    Buccal Cavity ... 11
    Esophagus ... 11
    Intestine ... 12
    Reproductive System ... 12
    Adhesion Tubes ... 12
    Cephalic Adhesion Tubes ... 12
    Posterior Adhesion Tubes ... 12
    Modified Adhesion Tubes ... 13
    Larval Adhesion Tubes ... 14
    Caudal Glands ... 14
Taxonomic History ... 14
Systematics ... 16
    Suborder Draconematina ... 16
    Superfamily Draconematoidea ... 16
    Family Draconematidae ... 17
    Subfamily Draconematinae ... 17
    Family Prochaetosomatidae n. fam. ... 18
    Subfamily Prochaetosomatinae n. subfam. ... 18
    Subfamily Cygnonematinae n. subfam. ... 18
    Subfamily Dracognominae n. subfam. ... 19
    Subfamily Notochaetosomatinae n. subfam. ... 19
    Superfamily Epsilonematoidea n. rank ... 19
Composition of the Suborder Draconematina ... 20
    Species Inquirendae of Draconematoidea ... 24
Key to Families, Subfamilies, and Genera of Draconematoidea ... 25
    Genus *Draconema* ... 27
        Larval Stages ... 27
        Key to Species ... 28

D. cephalatum	29
D. haswelli	30
D. claparedii	33
D. ophicephalum	36
D. antarcticum n. sp.	39
D. chilense n. sp.	40
Genus *Paradraconema* n. gen.	43
Larval Stages	45
Key to Species	46
P. floridense n. sp.	46
P. spinosum n. comb.	49
P. californicum n. sp.	50
P. newelli n. sp.	54
P. singaporense n. sp.	55
P. meridionale n. comb.	56
P. hopperi n. sp.	58
P. antarcticum n. sp.	62
Genus *Dracograllus* n. gen.	63
Larval Stages	65
Key to Species	66
D. cobbi n. sp.	66
D. mawsoni n. sp.	69
D. demani n. sp.	70
D. filipjevi n. sp.	72
D. timmi n. sp.	74
D. chitwoodi n. sp.	76
D. kreisi n. sp.	78
D. gerlachi n. sp.	79
D. stekhoveni n. sp.	81
D. wieseri n. sp.	84
D. eira n. comb.	87
D. falcatus n. comb.	87
D. solidus n. comb.	88
Genus *Dracotoranema* n. gen.	89
Larval Stages	89
D. trispinosum n. sp.	90
Genus *Prochaetosoma*	91
Larval Stages	91
Key to Species	93
P. primitiva	93
P. cayense n. sp.	94
P. vitielloi n. sp.	96
P. mediterranicum n. sp.	100
P. arcticum n. comb.	101
P. campbelli n. comb.	103
P. longicapitata n. comb.	104
P. lugubre n. comb.	105

Genus *Draconactus* n. gen.	105
Larval Stages	105
D. cutus n. sp.	106
D. suillus n. comb.	106
Genus *Apenodraconema* n. gen.	108
Larval Stages	108
Key to Species	108
A. chlidosis n. sp.	109
A. spinicaudum n. comb.	109
Genus *Cygnonema* n. gen.	111
Larval Stages	111
C. steineri n. sp.	112
Genus *Dracognomus* n. gen.	113
Larval Stages	113
Key to Species	115
D. marioni n. sp.	115
D. simplex n. comb.	118
D. notohalensis n. sp.	119
Genus *Notochaetosoma*	119
Larval Stages	121
N. tenax	121
Genus *Dracogalerus* n. gen.	124
Key to Species	124
D. afrikaanus n. sp.	124
D. bastiani n. sp.	125
D. cryptocephalus n. comb.	126
Summary	126
Literature Cited	131

Foreword

In its inception this monograph of the Draconematoidea represented the largest taxonomic endeavor of Dr. M. W. Allen's career. In its completion it is a work of love and a monument to a most distinguished scientific career. The work on this paper began in 1969 and at the time of his death in 1974 Dr. Allen had completed the 197 illustrations and selected the names for the proposed new superfamily, new family, four new subfamilies, eight new genera and 29 new species. The morphology and taxonomic history were in the state of a long hand draft. Ella Mae Noffsinger, who at that time was Dr. Allen's Staff Research Associate, had completed all specimen measurements. The task of bringing this monograph to completion unexpectedly became her responsibility. Miss Noffsinger has accomplished the task with a competency that is an attestment to Dr. Allen's teaching. The outstanding illustrations are an inspiration to all students of nematode taxonomy. This monograph is a model to be emulated.

A. R. Maggenti
Chairman, Division of Nematology
University of California, Davis

A REVISION OF THE MARINE NEMATODES OF THE SUPERFAMILY DRACONEMATOIDEA FILIPJEV, 1918 (NEMATODA: DRACONEMATINA)

by

M. W. Allen[1] and E. M. Noffsinger[2]

Acknowledgments

This study would not have been possible except for the many people who collected specimens throughout the world and furnished slides of type specimens for us to examine. We would like to thank the following for this invaluable service: W. D. Hope, U. S. National Museum, Smithsonian Institution, Washington, D. C.; A. M. Golden, USDA, Beltsville, Maryland; Elizabeth C. Pope, Australian Museum, Sydney, Australia; W. G. Inglis, South Australian Museum, Adelaide, Australia; S. Prudhoe, British Museum (Natural History), London, England; S. A. Gerlach, Institut für Meeresforschung, Bremerhaven, Germany; L. de Coninck and A. Coomans, Laboratoria voor Morfologie en Systematiek, Museum voor Dierkunde, Gent, Belgium.

We would especially like to thank the following workers and institutions for the numerous mass collections that were made available for this study. I. M. Newell, University of California, Riverside, furnished many of the collections used in this study. These collections were made possible through the support of the National Science Foundation: the deep sea material was collected during Cruise 17 of the ANTON BRUUN during the Southeastern Pacific Biological Oceanographic Program (1966); and the other material was collected during field work supported by NSF Grant GB 5027. Most of the Antarctica specimens used in this study were collected by R. W. Timm, C. O. R. R., Dacca, Bangladesh, and D. R. Viglierchio, University of California, Davis, on the project U. S. Antarctic Research Program (1969-1970) sponsored by the National Science Foundation. The collections from Port Jackson, Australia, type locality of Irwin-Smith's Australian species, were made by M. R. Sauer and P. Johnston, C. S. I. R. O., Merbein, Australia. P. Vitiello, Centre Universitaire de Luminy, Marseille, France, collected all of the specimens used from Marseille, France. The collections from Florida were provided by B. E. Hopper, Plant Quarantine Division, Agriculture Canada, Ottawa, Canada; G. C. Smart and Mr. DeNeve, University of Florida, Gainesville.

Our deep appreciation goes to P. Jensen, University of Copenhagen, Helsingør, Denmark, for not describing his new material in this superfamily prior to publication of this study.

E. Mae Noffsinger would like to thank all the members of the Division of Nematology, University of California, Davis, for their encouragement and valuable assistance in the completion of this manuscript. Deep appreciation goes to A. R. Maggenti for his valuable technical suggestions, reviewing the manuscript, and all the encouragement given me in this task. Special thanks also go to the following people for reviewing this manuscript: Virginia R. Ferris, Purdue University, Lafayette, Indiana; W. D. Hope, U. S. National Museum, Smithsonian Institution, Washington, D. C.; B. E. Hopper, Plant Quarantine Division, Agriculture Canada, Ottawa, Canada; and Hedwig H. Triantaphyllou, North Carolina State University, Raleigh. I would like to thank

---

1. The late Professor of Nematology, Department of Nematology, University of California, Davis 95616
2. Senior Museum Scientist, Division of Nematology, University of California, Davis 95616

Mrs. Merlin W. Allen for her assistance in the proofreading of this manuscript and her encouragement during its completion. Last, but not least, I wish to thank Mrs. Marjory McKenzie, Mrs. Roberta Grant, and Mrs. Ada Chater for typing the manuscript; and Mr. Tom Burlando for his assistance in proofreading.

Introduction

The suborder Draconematina (Desmodorida: Nematoda) as defined by de Coninck, 1965, contained a single superfamily Draconematoidea Filipjev, 1918, with 2 families, Draconematidae Cobb, 1929 and Epsilonematidae Steiner, 1927. We propose that the suborder is comprised of 2 superfamilies, Draconematoidea and Epsilonematoidea new rank. A diagnosis of the 2 superfamilies is presented in the Systematics. This suborder contains exclusively marine forms.

Draconematoidea are small nematodes, 0.3 to 3.0 mm in length. Distribution in marine waters is worldwide; specimens have been collected from all the oceans and many of the seas in tropical, temperate, arctic, and antarctic waters. Most specimens have been collected from depths of less than 50 meters, but 3 antarctic species have been collected from 540 meters. Improved collecting techniques, particularly the use of fine sieves and careful microscopic examination of recovered material, revealed draconematids to be more common and widely distributed than previous records indicated.

Since so little information is available about this group of nematodes, we have included the biological data available from our collections and the observations of previous workers. The highest number of species of Draconematoidea has been found in association with various kinds of marine algae. In many samples draconematids and epsilonematids are found together, and in association with the same algae, which may indicate that they have similar ecological requirements. In this study we found ectocommensals belonging to the Class Suctoria on some specimens of Draconematoidea.

Previous to this study, most new species descriptions have been based upon less than 6 specimens. This has resulted in an almost complete absence of information concerning intraspecific and generic variation of significant taxonomic characters. This revision is based upon the study of over 3,000 specimens from more than 80 locations.

The number of longitudinal rows and total number of posterior ventral adhesion tubes have been used as taxonomic characters to distinguish species. As the draconematids mature, the number of rows and total number of adhesion tubes increase with each life stage; in some previous descriptions the 2 rows of subventral adhesion tubes on adults have been mistaken for a single row. The numerous specimens examined in this revision enabled us to study all the life stages in several species and the variation in total numbers of posterior ventral adhesion tubes in most of the species.

Biology

Little is known about the biology of the Draconematoidea. The intestinal contents of the many specimens examined from all over the world gave no indication of their feeding habits. There have been no reports or studies conducted on environmental factors necessary for optimum reproduction.

In general, the preferred habitat seems to be intertidal; no specimens of this superfamily have been reported from open sea collections. Of the identified algae, draconematids were most commonly associated with red algae (table 1). It is not clear if these associations are primarily due to the availability of a particular food source, or the micro-environment afforded with these associations, or a combination of both.

Most suctorians (Phylum Protozoa, Subphylum Ciliphora, Class Suctoria) are ectocommensals, but a few are known to be parasitic, occurring in both salt and fresh water. Species belonging to 3 different genera of Suctoria (*Acineta* Ehrenberg, *Acinetopsis* Robin, and *Thecacineta* Collin) were present on specimens of Draconematoidea. The free-swimming young are ciliated, but the adults are without cilia and sessile. Suctorians attach themselves to smaller aquatic animals, plants, or inanimate objects. The relationship between suctorians and marine nematodes is not well defined. The genus *Draconema* had the highest incidence of association, both in the percentage of specimens and number of species with suctorians attached (table 2). Suctoria association does not appear to be related to nematode size, since many marine nematodes, besides the Draconematoidea, are known to bear suctorians. Draconematids were collected from depths ranging from shallow tide pools to 540 meters where the surface temperature of the water ranged from 40° to 80° F; so neither water depth nor temperature seems to be important. The surface texture of the cuticle and degree of nematode movement may be two factors involved in determining which nematodes become associated with Suctoria.

TABLE 1

Marine Habitat Associations of the Species of Draconematoidea

Nematode Species	Habitats*
Draconema	
D. cephalatum	1, 2, 3, 4, 5, 6, 9, 10, 11, 13, 14, 15, 16
D. haswelli	3
D. claparedii	1, 3
D. ophicephalum	1, 3
D. chilense n. sp.	14
D. antarcticum n. sp.	3
Paradraconema n. gen.	
P. floridense n. sp.	1, 3, 4, 5, 6, 9, 10, 12, 15
P. spinosum n. comb.	3, 16, 18
P. californicum n. sp.	11
P. newelli n. sp.	1, 5
P. singaporense n. sp.	7
P. meridionale n. comb.	1, 3, 5, 11, 16, 18
P. hopperi n. sp.	3, 9
P. antarcticum n. sp.	3
Dracograllus n. gen.	
D. cobbi n. sp.	9
D. mawsoni n. sp.	3
D. filipjevi n. sp.	11
D. demani n. sp.	3
D. timmi n. sp.	16
D. chitwoodi n. sp.	9
D. kreisi n. sp.	9

TABLE 1 (Cont.)

Marine Habitat Associations of the Species Draconematoidea

Nematode Species	Habitats*
D. gerlachi n. sp.	4
D. stekhoveni n. sp.	1, 3, 5, 16
D. wieseri n. sp.	8
D. eira n. comb.	3
D. falcatus n. comb.	17, 18
D. solidus n. comb.	3
Dracotoranema n. gen.	
D. trispinosum n. sp.	3
Prochaetosoma	
P. primitiva	3
P. cayense n. sp.	3, 9
P. vitielloi n. sp.	3
P. mediterranicum n. sp.	3
P. arcticum n. comb.	3
P. longicapitata n. comb.	14
P. lugubre n. comb.	3, 16
Draconactus n. gen.	
D. cutus n. sp.	1, 5, 15
D. suillus n. comb.	14
Apenodraconema n. gen.	
A. chlidosis n. sp.	1, 15
A. spinicaudum n. comb.	19
Cygnonema n. gen.	
C. steineri n. sp.	3
Dracognomus n. gen.	
D. marioni n. sp.	16
D. simplex n. comb.	3
D. notohalensis n. sp.	5
Notochaetosoma	
N. tenax	3
Dracogalerus n. gen.	
D. afrikaanus n. sp.	1, 3
D. bastiani n. sp.	1, 3
D. cryptocephalus n. comb.	3

*Habitats:
1. Algae
2. Barnacles
3. Bottom Debris
4. Brown Algae
5. Coral
6. Corallines
7. *Gracilaria* sp.
8. Green Algae
9. *Halimeda* sp.
10. Hydroid Colonies
11. Kelp Holdfasts
12. *Penicellus* sp.
13. *Phyllospadix* sp.
14. Red Algae
15. Red Corallines
16. Sand
17. Seaweed
18. Shells
19. *Tubipora* sp.

TABLE 2

Incidence of Suctoria in the Draconematoidea

Nematode	Locality	Adults (A) and/or larvae (L)	% Specimens affected	Number of species affected per genus
Draconema				
D. cephalatum	Colombia	A and L	3	5 out of 6
	Panama			
	Philippines			
	Society Islands			
D. haswelli	Australia	A and L	12	
D. claparedii	France	A and L	34	
D. ophicephalum	Italy	A and L	26	
D. antarcticum n. sp.	Antarctica	A and L	62	
Paradraconema n. gen.				
P. floridense n. sp.	Florida	A	3	2 out of 8
P. spinosum n. comb.	France	A and L	25	
Dracograllus n. gen.				
D. stekhoveni n. sp.	Colombia	A	2	1 out of 3
Dracotoranema n. gen.	France			0
Prochaetosoma				
P. cayense n. sp.	Florida	A	3	2 out of 8
P. vitielloi n. sp.	France	A and L	4	
Draconactus n. gen.	Panama			0
	Campbell Island			
Apenodraconema n. gen.	Society Islands			0
	Sarso Island			
Cygnonema n. gen.				
C. steineri n. sp.	Antarctica	A and L	6	1 out of 1
Dracognomus n. gen.	Florida			0
	France			
	Italy			
	Samoa Islands			
Notochaetosoma	Australia			0
Dracogalerus n. gen.	South Africa			0

Known and Recorded Distribution of Draconematoidea

ANTARCTIC OCEAN—*Fuegian Archipelago. Gauss Station. Ross Sea*—McMurdo Sound. Cape Royds and Scott Base, Ross Island. *Weddell Sea*—Duke Ernst Bay.

ARCTIC OCEAN—*Barents Sea. Svalbard.*

NORTH ATLANTIC OCEAN—*Baltic Sea*—Öresund. *Bay of Biscay. Black Sea*—Georgievskii Monastery, Kruglaya Bay and Sevastopol, Union of Soviet Socialist Republics. *Caribbean Sea*—Coco Solo and Galeta Beach, Panama. Jamaica, West Indies. *Danish Belt Sea. English Channel*—Roscoff and St. Vaast, France. *Greenland. Hudson Bay. Iceland*—Akureyri Harbor in Eyjafjörður Fjord. *Ireland*—Clare Island in Clew Bay. *Ligurian Sea*—Cala Cupa and Terrenia, Italy. *Mediterranean Sea*—Banyuls, Cannes, and Marseille, France. *North Sea*—Kiel Bay. *Norway. Scotland. Skagerrak. Tyrrhenian Sea*—

DISTRIBUTION OF DRACONEMATOIDEA

Bergen, Naples, Ischia Island and Salerno, Italy. *United States of America*—Bear Cut, Boca Raton, Coral Key, and Soldier Key, Florida. *Windward Islands*—Tobago.

SOUTH ATLANTIC OCEAN—*Brazil*—Ilha Bela, São Sebastiã Island. *Falkland Islands. Republic of South Africa*—Hout Bay. *Scotia Sea*—South Georgia Island.

INDIAN OCEAN—*Australia*—Christies Beach, Norman Ville Beach, and Port Noarlunga. *Kerguelen Island. Mascarene Islands. Red Sea*—Sarso Island. *Republic of South Africa*—East London.

SOUTH ATLANTIC OCEAN AND INDIAN OCEAN—*Republic of South Africa*—Cape Agulhas.

NORTH PACIFIC OCEAN—*Brasilan Strait*—Zamboanga, Mindanao, Mindanao Island, Philippine Islands. *Colombia*—Solano. *Gulf of California*—Puerto Libertad, Mexico. *Gulf of Panama*—Isla Taboga and Panama City, Panama. *Japan*—Akkeshi, Hokkaido Island. Nakaminato and Oarai, Ibaraki-ken; and Shimoda, Shizuoka-ken, Honshu Island. Ibusuki, Kyushu Island. *Mariana Islands*—Tamuning, Agana Bay, Guam Island. *Sea of Japan*—Wajima, Ishikawa-ken, Honshu Island, Japan. *Singapore Strait*—Changi Beach, Singapore Island. *South China Sea*—Little Santa Cruz Island and Matabungkay, Batangas, Luzon Island, Philippine Islands. *United States of America*—Dillon Beach, California. *Yellow Sea.*

SOUTH PACIFIC OCEAN—*Banda Sea*—Amboina Harbor, Amboina Island. Kei Island. *Campbell Island*—Persev. Harbor. *Chile*—Puerto Vagageria and Punta Caldera. *Coral Sea*—Nymphe Island, Australia. *Java Sea. Juan Fernandez Islands*—Mas A Tierra Island, Chile. *New Caledonia*—St. Vincent's Bay. *Samoa Islands*—Pago Pago Island, American Samoa. *Society Islands*—Motu Toopau, Bora Bora Island. Raiatea and Uturoa, Raiatea Island. Papeete, Tahiti Island. Tiarei, Tiarei Island. *Tasman Sea*—Broken Bay; Cremorne, Long Nose Point, Long Reef, and Vaucluse in Port Jackson, Australia.

EQUATOR-PACIFIC OCEAN—*Ecuador*—Punta Carnera. *Galapagos Islands*—Jensen Island, Santa Cruz Island, and Tower Island.

Abbreviations Used in Species Descriptions and Keys

ABD = Anal body diameter.
Acan-set = Acanthiform setae (see Morphology).
Ant = Anterior.
Ave = Average.
CAT = Cephalic adhesion tubes (dorsal side of rostrum). Number or length of CAT.
CAT (Ant) = Length of most anterior CAT (*Cygnonema* n. gen.).
CAT (Post) = Length of most posterior CAT (*Cygnonema* n. gen.).
Ceph = Cephalic region.
Ceph Acan-set = Length of largest sublateral acanthiform setae on rostrum, measured from base of seta to distal end.
Corn-set = Corniform setae (see Morphology).
first Mod-SlAT = Length of first posterior sublateral Mod-AT (most anterior sublateral tube, see Morphology).
first Mod-SvAT = Length of first posterior subventral Mod-AT (most anterior subventral tube).

first SlAT = Length of first posterior sublateral adhesion tube (most anterior sublateral tube).

first SvAT = Length of first posterior subventral adhesion tube (most anterior subventral tube).

first VAT = Length of first posterior adhesion tube in ventral row (most anterior tube).

Gub = Length of gubernaculum.

last Mod-SlAT = Length of last posterior sublateral Mod-AT (most posterior sublateral tube).

last SlAT = Length of last posterior sublateral adhesion tube (most posterior sublateral tube).

last SvAT = Length of last posterior subventral adhesion tube (most posterior subventral tube).

last VAT = Length of last posterior adhesion tube in ventral row (most posterior ventral tube).

Med Corn-set = (♂) Length of single large ventral corniform seta just anterior to first SvAT (*Dracotoranema* n. gen.).

Mod-AT = Modified adhesion tubes; small diameter, base of tubes slightly broadened, with distal end slightly expanded and open (*Dracognomus* n. gen. see Morphology).

Mod-CAT = Length or number of Mod-AT on rostrum (*Dracognomus* n. gen.).

Mod-SlAT = Posterior sublateral rows of Mod-AT.

Mod-SvAT = Posterior subventral rows of Mod-AT.

No = Number.

No Large SER Ann = Number of enlarged annules just posterior to rostrum (*Draconema*).

No Mod-SlAT = Number of Mod-AT in sublateral rows.

No Mod-SvAT = Number of Mod-AT in subventral rows.

Non-ann Term to Tail Length = Length of non-annulated tail region to total tail length (%).

No SlAT = Number of SlAT in sublateral rows.

No SvAT = Number of SvAT in subventral rows.

No VAT = Number of VAT in ventral row.

PAT = Posterior adhesion tubes (see Morphology). Number or length of SlAT and SvAT combined and given as range.

Post = Posterior.

Preanal Acan-set = (♂) Length of preanal acanthiform setae (measured from base of seta to distal end).

Preanal Corn-set = (♂) Length of preanal corniform setae (measured from base of seta to distal end).

PS = Somatic setae borne on conspicuous, raised, cuticular pedicels (*Dracograllus* n. gen.).

SER = Swollen esophageal region.

SER (E/L) = Length of swollen body area in esophageal region, expressed as % of body length.

SER (L/W) = Length of swollen body area in esophageal region divided by greatest body width in this area.

SlAT = Sublateral rows of posterior adhesion tubes. Length of first and last tubes sometimes given as range.
Spic = Length of spicule.
SS = Somatic setae (see Explanation of Measurements).
SvAT = Subventral rows of posterior adhesion tubes. Length of first and last tubes sometimes given as range.
T/ABD = Total tail length divided by anal body diameter.
Vent = Ventral.
VAT = Posterior adhesion tubes in single ventral row.

Explanation of Measurements Used in Species Descriptions and Keys

1. No *a* measurement is given for females because most are swollen, and this measurement is not a useful taxonomic character.

2. No gonad lengths are given for either sex; these measurements are not useful taxonomic characters.

3. Most measurements and counts of posterior adhesion tubes have been made on the right side of the nematodes.

4. All μm measurements and averages of any structures counted have been rounded off to nearest whole number.

5. Measurements of swollen esophageal region were taken as follows:
    a. *Length*—measured from the anterior tip of the lip region to just posterior to the swollen esophageal region; in most of the Draconematidae the body is constricted at this region.
    b. *Width*—the body diameter was measured at the widest part of the swollen esophageal region.

6. Position of setae and cuticularized protuberants on the non-annulated tail region was measured from the last complete tail annule to tail tip, and expressed as a percent.

7. Location of cephalic adhesion tubes when posterior to rostrum, expressed as rostral widths posterior to rostrum.

8. Rostral width was measured at the base of the rostrum just anterior to first body annule.

9. Somatic setae measurements expressed as a range, only the *longest* setae were measured on the rostrum, swollen esophageal region, opposite PAT, and annulated tail region.

Morphology

Body shape of draconematids is variable but generally is a modification of cylindrical. In some genera the markedly swollen esophageal region is followed by a narrowed cylindrical portion that gradually increases in diameter to mid-body. The mid-body of females, at the level of the gonads, has a diameter equal to or greater than that of the esophageal region. Males of some species show a similar increase in mid-body diameter (fig. 176). In both sexes the body is generally cylindrical from mid-body to anus. The caudal region in both sexes is conoid to elongate-cylindrical. When relaxed the body curvature is generally sigmoid. Some species have a body shape resembling that of the epsilonematids. General body shape is well adapted to the "inchworm" type of movement that has been observed to be characteristic of live specimens.

Cuticle.

The cuticle has conspicuous annulations extending from the non-annulated cephalic region (rostrum) to the caudal region. In the non-annulated cephalic region the cuticle is strongly thickened forming a heavily cuticularized structure called the helmet, but the cuticle of the labial region is not thickened and remains pliable. Length of the non-annulated, punctated terminal portion of the tail is variable in the superfamily. In some species the annules immediately posterior to the rostrum differ from subsequent annules by being larger and more heavily cuticularized. These larger annules are usually ornamented by subcuticular markings of vacuolar type (Chitwood and Chitwood, 1937).

External cuticular ornamentation is not common; but sometimes there are fringes or minute spines at the margins of the annules, or less frequently annular ridges with spine-like projections (fig. 166). Annules on some species may be longitudinally areolated (fig. 75). Subsurface cuticular markings may also be present on the rostrum. The non-annulated tail terminus is ornamented with conspicuous or inconspicuous punctations.

Setae.

Somatic setae occur along the entire length of the body. On the annulated portion of the body in adults they are usually arranged in 8 longitudinal rows, 2 subdorsal, 4 sublateral and 2 subventral; the number of rows is variable in the larval stages. Somatic setae (not in rows) are also present on the rostrum and non-annulated region of the tail. In some species conspicuous long and short somatic setae alternate in each row. In *Dracograllus* n. gen. some species have somatic setae borne on conspicuous, raised, cuticular pedicels (pedicel-setae, fig. 110).

Most males have paired anal setae located anteriorly and posteriorly and subventral or sublateral to the anus (cloacal opening). These setae are generally uniformly tapered (fig. 24); but in 2 species of *Draconema* some of the anal setae are short, have a round broad-base and narrow abruptly near the mid-region (unevenly tapered, figs. 5, 48). Many of the females have 2 pairs of subventral paravulval setae, 1 pair anterior and 1 posterior to the vulva (figs. 22-23).

Some species have acanthiform, thorn-shaped (Acan-set), or corniform, horn-shaped (Corn-set) setae. These setae have a "spine-like" appearance, but since they are innervated, with the base extending into the body, they cannot be spines, but are actually modified setae. Acanthiform setae occur on the rostrum in both sexes (fig. 94), but occur as preanal setae only in males (fig. 67). Corniform setae (fig. 146) have been observed only on males as preanal and mid-body setae.

Lips.

The lips and that portion of the cephalic region anterior to the heavily cuticularized helmet are covered by a thin cuticle. In fixed specimens the lips may be withdrawn into the anterior part of the helmet.

In Draconematidae there are 6 well-developed lips, each bearing a labial sensory seta.

In the original description of *Draconema cephalatum* in 1913, Cobb reported the presence of rib-like structures within the lip region. This internal cuticularized framework consists of 12 outwardly projecting ribs (fig. 3). Their position conforms to the sectors of the esophagus; they are arranged in a dorsal series of 4 ribs, and 2 subventral series each with 4 ribs. Each rib appears to be composed of two parts, a medial part lying parallel to the stoma; and a curved part, the rib, extending posteriorly and radially (fig. 9). These ribs are very similar to, and may well be homologous with the 12 projectable cheilostomatal rugae found in Chromadorida.

In Prochaetosomatidae the lips are obscure and their presence is evidenced by the internal circle of sensory organs which are either papilloid or very short inconspicuously setose. There is an internal framework below the lip region that is not well developed in most genera.

Cephalic Sensory Organs.

The cephalic sensory organs appear to have a 6 + (6 + 4) arrangement. However, in most genera the presence of numerous cervical setae obscures the symmetry of the outer circles.

Amphids.

The amphids are elongate loop-shaped to spiral. In the Draconematidae the amphids are lateral on the rostrum, males usually have loop-shaped amphids (fig. 57), and females have spiral (fig. 58) or loop-shaped amphids. Prochaetosomatidae exhibit more variation in the shape of the amphid, which essentially represents modifications of the loop-shaped or spiral, frequently located dorso-lateral on the rostrum. *Dracognomus* n. gen. has small obscure amphids that are inverted "U" shaped, staple-shaped (fig. 178).

Amphids in the genus *Draconema* are large and conspicuous. Because of this we attempted to study their external structure using the scanning electron microscope. The procedures (Sher and Bell, 1975; and Norman Jones, University of California, Davis, personal communication) used to prepare specimens were not satisfactory; because they caused delicate membranous structures to collapse, such as the distal ends of the setae, and the cephalic and posterior adhesion tubes. In our preparations the amphids appear as grooves in the rostrum. In scanning electron micrographs, the dorsal arm of the groove appears as a hole where it penetrates the rostrum (figs. 198, 199). We believe that there is a delicate membranous tube that lies in the amphidial groove, and that this tube is collapsed in our preparations. In support of this conclusion, we have observed in some of the specimens in *Paradraconema* n. gen. that were fixed in 70% alcohol, a separation of the amphidial tube from the groove in which it normally lies. Two-thirds of the tube was completely out of the groove with the proximal $^1/_3$ still lying in the groove to the point where the dorsal arm of the groove extends internally (fig. 93). The membranous tube is circular in cross-section when viewed under the light microscope. The distal end of the detached tube appears to be open but this may be the result of separation from the groove. Riemann (1972) believes this tube is filled with a gelatinous substance called "corpus gelatum".

Buccal Cavity.

The stoma in Draconematidae is weakly developed, without armature, and not distinguishable in toto mounts from the anterior lumen of the esophagus (fig. 2). In Prochaetosomatidae, the stoma, excepting *Notochaetosoma,* is weak to moderately well developed, armed with a dorsal tooth and sometimes with small obscure subventral teeth. The buccal cavity may be shallow and partially collapsed (fig. 165) or elongate conical with a wider chamber anterior to the dorsal tooth (fig. 157).

Esophagus.

The three-part esophagus in Draconematidae consists of anterior and posterior non-valved swellings separated by a constricted isthmus encircled by the nerve ring (fig. 49). In draconematids there is a conspicuous esophago-intestinal valve. The majority of the Prochaetosomatidae have a two-part esophagus consisting of a cylindrical corpus and a terminal bulb, sometimes with a cuticularized valve (fig. 148). The esophago-intestinal valve is small and obscure in prochaetosomatids.

Intestine.

The intestine has relatively few cells, 6 to 8 in cross-section. It is a relatively straight tube, except where displaced by the gonads.

Reproductive System.

The female has a didelphic, amphidelphic reproductive system, with short, club-shaped, reflexed ovaries. The vulva is located near mid-body. The male has a monorchic, anteriorly directed system, usually without flexures. There are 2 arcuate cephalated spicules, and a simple gubernaculum which is trough-shaped at the distal end.

Males in some species of Draconematidae have a single ventral or paired sublateral preanal acanthiform setae (Acan-set); or large paired or unpaired, preanal corniform setae and/or in combination with a large ventral corniform seta (Corn-set) just anterior to the posterior adhesion tubes. Prochaetosomatidae males do not have supplementary acanthiform or corniform setae.

Adhesion Tubes.

These specialized structures are hollow, and associated with glands (Irwin-Smith, 1918; Cobb, 1929) which are presumed to secrete a material that attaches the tubes to the substrate allowing the nematode to move in an "inchworm-like" manner. These tubular structures have been referred to as tubular setae or ambulatory setae in the literature (Kreis, 1938; de Coninck, 1965). Since the structures are tubular and not setiform, we are following the nomenclature proposed by Cobb (1929) and call these structures adhesion tubes. The "stilt setae" that are found in Epsilonematoidea differ from the "adhesion tubes" of Draconematoidea by being solid and not known to be associated with glands.

Cephalic Adhesion Tubes (CAT).

These adhesion tubes are located on the dorsal sector of the cephalic region (fig. 21). Irwin-Smith (1918) and Cobb (1929) believed the CAT enables draconematids to draw the posterior part of the body forward during their "inchworm-like" movement.

The base of the CAT is normally broad, and then the tube gradually tapers to the slightly expanded and open distal end. Cobb (1929) indicates that they have the same internal structure at the distal end as the posterior adhesion tubes; however, we have not observed any evidence of the presence of the characteristic tongue-like structure that is found in the distal end of the posterior adhesion tubes.

There is considerable variation in the position of the CAT, but all of the tubes are located dorsally on the rostrum. In Draconematidae they are generally in 2 transverse rows consisting of 4 or 6 pairs. In a few species there is a total of 14 or 15 tubes, in these instances, there are 3 tubes rather than 1 pair in the series nearest the amphid.

In Prochaetosomatidae the CAT may be located on the rostrum or on the annulated region just posterior to the rostrum. They may be in a single transverse row, but more commonly they are in 2 transverse rows of paired tubes. In *Cygnonema* n. gen. they occur in 3 or more transverse rows. Number of tubes in prochaetosomatids varies from 4 to 14.

Posterior Adhesion Tubes (PAT).
These tubular structures are located on the ventral sector of the posterior half of the body, and are designated as sublateral adhesion tubes (SlAT) and subventral adhesion tubes (SvAT). Cobb's (1913) description of *D. cephalatum* is very detailed and for the first time notes the presence of 4 rows of adhesion tubes in an adult. All adult draconematids possess at least 4 longitudinal rows, 2 sublateral (SlAT) and 2 subventral (SvAT). In the genus *Prochaetosoma* 1 species exhibits a modification of the 4 row condition by having 4 rows (2 sublateral, 2 subventral) in the anterior region and 3 rows (2 sublateral, 1 ventral) in the posterior region (fig. 154). A single female of an undescribed species collected in the Mediterranean Sea off the coast of France has 6 rows in the anterior $1/3$ and 4 rows in the posterior $2/3$. The number and length of the adhesion tubes is sometimes useful for separation of species.

Each adhesion tube terminates in a bell-shaped enlargement which is open at the distal end (fig. 175). The cuticle of the enlargement is thin and membranous in contrast to the thicker cuticle of the cylindrical portion of the tube. The thick cuticle probably contributes a structural rigidity to the tube. The bell-shaped portion contains a tongue-like triangular structure that extends from the base of the bell towards the open distal end. Cobb (1929) suggested that this structure represented an extension of the lumen of the tube into the bell. However, the lumen of the tube empties into the bell on the ventral side of the tongue-like extension (fig. 175).

The adhesion tubes are associated with a longitudinal series of internal glands, usually 9 in number, which are believed to secrete a material that is used to attach the tubes to the substrate (fig. 160). Cobb (1929) suggested that the tubes, because of their arrangement in 4 rows with a difference in length between the outer and inner rows, are used to attach the nematode to a filamentous substrate such as marine algae.

The SlAT and SvAT rows of adhesion tubes are situated in positions corresponding to the longitudinal rows of somatic setae, which has led to speculation that the tubes, both cephalic and posterior, might be modified somatic setae. Filipjev in 1921 rejected Steiner's 1916 proposal that the CAT were modified setae. Females have short somatic setae intermingled with the SlAT, and the number of these setae is the same as that of the somatic setae in the adjacent sublateral row (figs. 51-52). In males the same distribution of somatic setae is present within the rows of SlAT, but some of the setae are long and not reduced in length. Distribution of these long setae is frequently useful in the separation of species. Usually no setae occur within the rows of SvAT at the anterior end of the row, but a few short setae are usually present near the posterior end.

In *Dracogalerus* n. gen., the males of some species have corniform adhesion tubes at the posterior end of the SvAT (fig. 196). We have used the term corniform because of their shape and heavy cuticularization. These tubes exhibit a gradual change in the bell and tongue-like structure, until the most posterior tube appears to lack this structure.

Modified Adhesion Tubes (Mod-AT).
Mod-AT are present on species in *Dracognomus* n. gen. These adhesion tubes are smaller in diameter than the normal adhesion tubes. They are hollow; their bases are only slightly expanded; and the tube gradually tapers to the distal end which is slightly expanded and open. The internal structure of the distal end is obscure, and the tongue-like extension was not observed (fig. 177). Both types of adhesion tubes, Mod-AT and normal adhesion tubes, may be present (fig. 179). In addition to the normal location of the cephalic and posterior adhesion tubes, some of the species have Mod-AT located ventrally on the esophageal and/or mid-body regions.

Larval Adhesion Tubes.
The general position and morphology of the cephalic adhesion tubes (CAT) and posterior adhesion tubes (PAT) in larval stages are similar to those found in the adults. Larval tubes differ from the adult tubes by size and number. The number of CAT, and number of longitudinal rows of PAT and total number of tubes differentiates the larval stages.

No adhesion tubes occur on the first larval stage (figs. 33, 38); both the CAT and PAT are absent.

In the many second-stage larvae examined the adhesion tubes did not vary in number. This stage has a single CAT, and 2 pairs of sublateral adhesion tubes (SlAT) in 2 longitudinal rows (figs. 34, 39).

The total number of adhesion tubes in the third and fourth larval stages depends upon the total number of tubes present in the adults. All third-stage larvae examined had 2 longitudinal rows of SlAT. *Draconema* third-stage larvae have 3 CAT, and 3 pairs of SlAT (figs. 35, 40).

Most fourth-stage larvae have 4 CAT; and 3 longitudinal rows of PAT, 2 sublateral and 1 ventral (figs. 37, 41). In 3 of the genera the number of CAT ranges from 4 to 8 tubes. In all known genera of this superfamily having all normal adhesion tubes the PAT occur in 3 longitudinal rows on fourth-stage larvae, but *Dracognomus* n. gen. has modified and/or normal adhesion tubes and these tubes occur in 4 rows (2 sublateral and 2 subventral).

Caudal Glands.
In all species examined there are 3 glands arranged in tandem in a dorsal position in the posterior part of the body. The glands extend anterior to the anus, and usually open through the spinneret at the tail terminus. In 2 species of *Prochaetosoma* there appears to be no spinneret or orifice and the glands appear abnormal (figs. 155, 159).

Taxonomic History

Most of the confusion in this superfamily has resulted from the descriptions of immature forms or the inaccurate observation of the number of longitudinal rows of the posterior adhesion tubes. After examining over 3,000 specimens in this superfamily, we have found that the adult specimens always have at least 4 longitudinal rows of posterior adhesion tubes, fourth-stage larvae have 3 rows (except *Dracognomus* n. gen., see Morphology, p. 14), and the third- and second-stage larvae have only 2 rows.

Chaetosoma ophicephalum was described by Claparéde in 1863; in the intervening century the generic names *Tristicochaeta* Panceri, 1876; *Draconema* Cobb, 1913;

Drepanonema Cobb, 1933; and *Claparediella* Filipjev, 1934 have been used for forms that we believe are congeneric. The generic name *Chaetosoma* Claparéde, 1863 was preoccupied by *Chaetosoma* Westwood, 1851, a coleopteron. *Chaetosoma ophicephalum* Claparéde, 1863 was described from an adult female characterized as having 2 rows of subventral adhesion tubes and illustrated as having an enlarged posterior esophageal bulb and a smaller anterior swelling. *Tristicochaeta inarimense* was first proposed by Panceri in 1876 and was followed in 1878 with a more detailed description and illustrations. Panceri proposed the new generic name because his specimens had 2 and 3 longitudinal rows of posterior ventrally located adhesion tubes. The specimens described by Panceri were not adults. In 1907, Schepotieff, not being aware that the generic name *Chaetosoma* was preoccupied, synonymized *Tristicochaeta* with *Chaetosoma*. Cobb (1913) proposed *Draconema cephalatum* without reference to prior publications and without indicating how *Draconema* differed from *Chaetosoma* or *Tristicochaeta*. Filipjev (1918) recognized *Tristicochaeta* as a valid name. He further expressed his belief that *Draconema* was not so different from *Chaetosoma* as to warrant a separate genus, but recognized *Draconema* as the valid name because *Chaetosoma* was preoccupied. In this decision we concur.

In 1929, Cobb proposed that *Notochaetosoma* Irwin-Smith, 1918, and not *Draconema*, replace the preoccupied name *Chaetosoma*. A posthumous paper edited by Margaret V. Cobb (1933) credited N. A. Cobb with a previously unpublished new generic name *Drepanonema* to replace *Chaetosoma;* this was listed in a chart. In 1934, Filipjev proposed the new name *Claparediella* to replace *Chaetosoma* and recognized *Draconema* and *Notochaetosoma* as valid names. In a key posthumously published by Margaret V. Cobb and Corinne Cooper (Cobb, 1935), N. A. Cobb is credited with recognizing *Draconema* as a synonym of *Tristicochaeta* and re-emphasized that *Drepanonema* is the valid name of *Chaetosoma*. This key was developed and used by N. A. Cobb as a card catalogue during his 40 years in nematology.

In 1938 Johnston reviewed the status of the generic names discussed in the foregoing account. He followed Cobb's (1935) proposal that *Drepanonema, Tristicochaeta* and *Notochaetosoma* were the valid names and proposed Drepanonematidae as a new family name; however, Cobb (1929) had already proposed the family name Draconematidae.

The most recent taxonomic reviews of nematodes in the family are those of de Coninck, 1965; Hope and Murphy, 1972; and Gerlach and Riemann, 1973. De Coninck recognized *Draconema* Cobb, 1913; *Drepanonema* Cobb, 1933 (syn. *Chaetosoma* Claparéde, 1863 and *Claparediella* Filipjev, 1934); *Notochaetosoma* Irwin-Smith, 1918 and *Tristicochaeta* Panceri, 1876 as valid names. Hope and Murphy recognize *Draconema* Cobb, 1913 (syn. *Chaetosoma* Claparéde, 1863; *Claparediella* Filipjev, 1934; and *Drepanonema* Cobb, 1933); *Notochaetosoma* Irwin-Smith, 1918; and *Tristicochaeta* Panceri, 1876 as valid names. Gerlach and Riemann recognize *Draconema* Cobb, 1913; *Drepanonema* Cobb, 1933 (syn. *Chaetosoma* Claparéde, 1863; *Claparediella* Filipjev, 1934; and *Tristicochaeta* Panceri, 1876); *Notochaetosoma* Irwin-Smith, 1918; and *Prochaetosoma* Micoletzky, 1922 as valid names. Gerlach and Riemann also stated that the name *Draconema* should be preserved, as it is the only one used in publications.

Since the genus *Chaetosoma* is preoccupied, the next genus to replace it would be

Tristicochaeta. Panceri proposed *T. inarimense*, new genus and new species because his specimens had 2 and 3 longitudinal rows of posterior adhesion tubes. He distinguished this species from *C. ophicephalum* Claparéde, 1863 (now *Draconema ophicephalum*) on the basis of the 3 longitudinal rows of posterior adhesion tubes. From our studies of this superfamily, the specimen with 2 longitudinal rows of 2 pairs of sublateral adhesion tubes in each row was a second-stage larva, and the specimen with 3 rows (2 sublateral and 1 ventral) was a fourth-stage larva.

Tristicochaeta inarimense is in the Draconematidae because of the definite three-part esophagus and swollen esophageal region which was illustrated by Panceri. The fourth-stage larva of this species was described as having 3 longitudinal rows of posterior adhesion tubes composed of 5 pairs of sublateral adhesion tubes and 8 tubes in the ventral row. Our examinations of over 500 fourth-stage specimens in Draconematidae indicate that the number of sublateral adhesion tubes varies from 5 to 7 pairs, and from 5 to 13 tubes in the ventral row. Among these fourth-stage larvae were specimens collected from Naples, Italy, and all of these had 5 pairs of sublateral adhesion tubes and 9 tubes in the ventral row. There was no indication either in the description nor any indication in the illustrations of the amphid, cephalic adhesion tubes, presence or absence of enlarged annules just posterior to the rostrum, or cephalic acanthiform setae.

Because *T. inarimense* was described from immature specimens in which the only major taxonomic characters given in the description and illustrations overlap throughout Draconematidae; and because all the specimens we examined from near the type locality did not fit the description, we have had to conclude that this species is unidentifiable. Therefore *T. inarimense* is made a *Species Inquirenda* and the next valid generic name to replace *Chaetosoma* is *Draconema.*

Systematics

SUBORDER Draconematina de Coninck, 1965

Diagnosis: Desmodorida. Helmet present. Lips obscure to well developed. Cephalic sensory setae frequently interspersed with cervical setae. Amphids spiral to elongate loops. Cephalic adhesion tubes present or absent. Cuticle annulated, sometimes ornamented with spines, ridges, granules, or internal vacuoles. Somatic setae present. Posterior adhesion tubes or stilt setae usually present. Female amphidelphic, didelphic, ovaries reflexed, vulva near mid-body. Male monorchic, testis outstretched. Two curved, cephalated spicules; gubernaculum present.

Two superfamilies: Draconematoidea Filipjev, 1918 and Epsilonematoidea n. rank.

SUPERFAMILY Draconematoidea Filipjev, 1918[3]

Diagnosis Emended: Draconematina. Variable in length, 0.3 to 3.0 mm. Relaxed body configuration usually dorsally and ventrally arched into shallow sigmoid shape. Greatest body width usually in esophageal and mid-body regions. Cephalic sensory organs 6 + (6 + 4); inner circle papilloid or setose; outer circles setose, obscured by numerous cervical setae. Amphids elongate loop-shaped to spiral, located lateral to dorso-lateral on non-annulated cephalic region (rostrum). Lips obscure to conspicuous, supported by an internal cuticularized framework. Cephalic adhesion tubes (CAT)

---

3. Authorship of family-group names according to the International Code of Zoological Nomenclature, Article 36.

present, located dorsally on rostrum or just posterior to rostrum. Buccal cavity obscure to moderately well developed, armed or unarmed. Esophagus either with anterior and posterior swellings separated by constriction or with cylindrical corpus and posterior bulb. Posterior bulb with or without valve. Cuticle with conspicuous annules, except for rostrum and tail terminus, annules sometimes heavily cuticularized; rarely ornamented with spines, ridges or internal vacuoles. Adhesion tubes hollow, glandular. Stilt setae absent. Posterior adhesion tubes present, in 4 or more longitudinal rows in posterior $1/3$ of body. Sometimes rows of SlAT extend posterior to anus. Gubernaculum simple, trough-shaped at distal end. Sex ratio almost equal. Three caudal glands present, extending anterior to anus; usually with spinneret. Tails of both sexes conoid to elongate-cylindrical, with non-annulated terminal region ornamented with punctations.

Two families: Draconematidae Filipjev, 1918 and Prochaetosomatidae n. fam.

FAMILY Draconematidae Filipjev, 1918

Syn: Chaetosomatidae Schepotieff, 1907
Drepanonematidae Johnston, 1938
Claparediellidae Allgén, 1954

Diagnosis Emended: Draconematoidea. Esophageal region of body swollen, followed by reduced body width, enlarging gradually to mid-body. Females greatest body width at mid-body, males usually at esophageal region. Cephalic sensory organs setose. Six conspicuous lips, with internal supporting framework consisting of 12 rib-like rods. Amphids loop-shaped to spiral, lateral on rostrum. CAT on rostrum. Sublateral cephalic acanthiform setae sometimes present on rostrum. Buccal cavity inconspicuous, collapsed, unarmed. Esophagus with enlarged corpus and posterior bulb, separated by short isthmus surrounded by nerve ring, esophagus "dumbbell-shaped". Posterior bulb without valve. Sometimes with larger, heavily cuticularized annules just posterior to rostrum. Annules without ornamentation except very minute spines in some species. SS in 8 longitudinal rows along entire body length to anus, tail with 4 sublateral rows. SS sometimes borne on conspicuous, raised, cuticular pedicels. Posterior adhesion tubes in 4 longitudinal rows, some tubes may be posterior to anus. Males of some species with supplementary preanal acanthiform or corniform setae, or combination of mid-body and preanal corniform setae. Caudal glands present, with spinneret. Tails of both sexes cylindrical-conoid.

Discussion.—(See Taxonomic History.) In 1907 Schepotieff proposed the family Chaetosomatidae (= Chaetosomatiden), not being aware the name was preoccupied by Chaetosomatidae Shipley, 1896 (insect); in this family he included the genus *Chaetosoma* Claparéde, 1863. Johnston (1938) proposed a new family name Drepanonematidae for *Drepanonema* Cobb, 1933, *Tristicochaeta* Panceri, 1876 and *Notochaetosoma* Irwin-Smith, 1918. Allgén (1954) proposed the new family name Claparediellidae for *Claparediella* Filipjev, 1934. Neither of these family names is valid because Cobb (1929) had already proposed the family name Draconematidae based on the type genus *Draconema* Cobb, 1913, which currently has 1 subfamily.

SUBFAMILY Draconematinae Filipjev, 1918

Diagnosis: Draconematidae. Since only 1 subfamily is recognized, the characters of the family also serve as the subfamily diagnosis.
Type Genus: *Draconema* Cobb, 1913
Other Genera: *Paradraconema* n. gen.
Dracograllus n. gen.
Dracotoranema n. gen.

Discussion.—Filipjev in 1918 proposed the subfamily name Draconematinae (= Draconematini), which included *Draconema* Cobb, 1913 and *Tristicochaeta* Panceri, 1876.

FAMILY Prochaetosomatidae n. fam.

Diagnosis: Draconematoidea. Esophageal region of body not swollen or only slightly swollen. Greatest body width at mid-body or in region of posterior adhesion tubes. Inner circle of cephalic sensory organs usually papilloid, outer circles setose. Six inconspicuous lips, with obscure supporting framework. Amphids loop-shaped to spiral, usually dorso-lateral on rostrum. CAT posterior to or on rostrum. Buccal cavity usually moderately developed and armed with dorsal tooth, sometimes collapsed and unarmed. Esophagus with cylindrical corpus and terminal swelling or bulb with or without cuticularized valve. SS in 6 to 8 (usually 8) longitudinal rows along body to anus, tail with 4 sublateral rows. Posterior adhesion tubes in at least 4 longitudinal rows in anterior region of rows, some tubes may be posterior to anus. Caudal glands present, usually with spinneret. Tails of both sexes conoid to elongate-cylindrical.

Four subfamilies: Prochaetosomatinae n. subfam.
Cygnonematinae n. subfam.
Dracognominae n. subfam.
Notochaetosomatinae n. subfam.

SUBFAMILY Prochaetosomatinae n. subfam.

Diagnosis: Prochaetosomatidae. Anterior 25% to 45% of body slender, with or without conspicuous swollen regions. Rostrum rounded, conical or broadly flattened anteriorly. Amphids on posterior half or near center of rostrum, conspicuous, elongate loop-shaped, loop-shaped (horseshoe- or crescent-shaped), circular or elongate unispiral, double spiral. CAT in 1 to 3 transverse rows usually posterior to rostrum, posterior row 1 or less rostral widths (4 to 11 annules) posterior to rostrum. Stoma opening near median line. Buccal cavity weak (partially collapsed) to moderately developed, with conspicuous or inconspicuous dorsal tooth; with or without (usually without) 1 to 2 minute, inconspicuous, ventrally located teeth. Esophagus short with cylindrical corpus, posterior bulb with or without cuticularized valve. Cuticle not unusually thick. Annules with or without ornamentation. Modified adhesion tubes (Mod-AT) absent. Posterior adhesion tubes in 4 longitudinal rows or with 4 rows in anterior region and 3 posterior rows, no tubes posterior to anus. Spinneret present or absent. Tails of both sexes conoid to cylindrical-conoid, or elongate cylindrical-conoid (spike-like).

Type Genus: *Prochaetosoma* Micoletzky, 1922
Other Genera: *Draconactus* n. gen.
Apenodraconema n. gen.

Differs from Cygnonematinae by anterior 25% to 45% of body slender; CAT in 1 to 3 transverse rows, posterior row 1 or less rostral widths (4 to 11 annules) posterior to rostrum. Differs from Dracognominae by absence of modified adhesion tubes (Mod-AT) and median swelling in esophagus. Differs from Notochaetosomatinae by presence of a posterior esophageal bulb (with or without cuticularized valve) and dorsal tooth in buccal cavity.

SUBFAMILY Cygnonematinae n. subfam.

Diagnosis: Prochaetosomatidae. Anterior 50% to 59% of body long and slender, without conspicuous swollen areas. Rostrum broadly rounded anteriorly. Amphids on posterior half of rostrum; male amphids conspicuous, elongate loop-shaped; female amphids inconspicuous, tubular-shaped (single elongated tube). CAT in 3 or more transverse rows usually posterior to rostrum, posterior row 2 to 3 rostral widths (26 to 37 annules) posterior to rostrum. Stoma opening near median line. Buccal cavity weakly developed, with conspicuous dorsal tooth. Esophagus long with elongate cylindrical corpus, and terminal swelling without cuticularized valve. Cuticle not unusually thick. Annules with obscure ornamentation. Modified adhesion tubes (Mod-AT) absent. Posterior adhesion tubes in 4 longitudinal rows, no tubes posterior to anus. Spinneret present. Tail of both sexes cylindrical-conoid.

Type Genus: *Cygnonema* n. gen.

Differs from other subfamilies in Prochaetosomatidae by anterior 50% to 59% of body long and slender; long esophagus; CAT in 3 or more transverse rows, with posterior row 2 to 3 rostral widths (26 to 37 annules) posterior to rostrum.

SUBFAMILY Dracognominae n. subfam.

Diagnosis: Prochaetosomatidae. Esophageal region of body slightly swollen, followed by slender region, mid-body slightly to obviously swollen. Rostrum ventrally directed, dorsal side oblique, with anterior broadly rounded. Amphids inconspicuous, inverted "U" shape (staple-shaped), at base of rostrum extending through first few body annules. Mod-CAT in 2 transverse rows on anterior half of rostrum or rows less than 1 rostral width (3 to 4 annules) posterior to rostrum. Stoma opening slightly ventral. Buccal cavity well developed, cylindrical, with conspicuous dorsal tooth. Esophagus short, corpus with small median swelling, posterior bulb with conspicuous cuticularized valve. Cuticle not unusually thick. Annules with or without ornamentation. Modified adhesion tubes (Mod-AT) present, small diameter, hollow, bases slightly expanded, tubes gradually tapered to distal end, distal end slightly expanded and open with internal structure obscure. Esophageal and mid-body regions with or without Mod-AT. Normal adhesion tubes present or absent. Posterior adhesion tubes in 4 longitudinal rows, with or without tubes posterior to anus. Spinneret present. Tail of both sexes cylindrical-conoid.

Type Genus: *Dracognomus* n. gen.

Differs from other subfamilies in Prochaetosomatidae by inverted "U" shaped (staple-shaped) amphid, presence of modified adhesion tubes (Mod-AT), dorsally oblique rostrum, and esophagus with small median swelling.

SUBFAMILY Notochaetosomatinae n. subfam.

Diagnosis: Prochaetosomatidae. Body width nearly equal throughout length of body. Rostrum conoid or broadly rounded anteriorly. Amphids on anterior half of rostrum, conspicuous, loop-shaped (horseshoe-shaped), unispiral, or partially doubled spiral. CAT in 2 to 3 transverse rows on anterior half of rostrum, or with anterior row on rostral base and posterior row less than 1 rostral width (2 to 3 annules) posterior to rostrum. Stoma opening near median line. Buccal cavity collapsed or weak to moderately developed, unarmed. Esophagus short with cylindrical corpus, inconspicuous terminal swelling without cuticularized valve. Cuticle very thick; thickness may obscure internal structures. Annules with or without ornamentation. Modified adhesion tubes (Mod-AT) absent. Posterior adhesion tubes in 4 longitudinal rows, no tubes posterior to anus. Spinneret present. Tail of both sexes cylindrical-conoid.

Type Genus: *Notochaetosoma* Irwin-Smith, 1918
Other Genus: *Dracogalerus* n. gen.

Differs from other subfamilies in Prochaetosomatidae by very thick cuticle, and absence of a dorsal tooth in buccal cavity.

SUPERFAMILY Epsilonematoidea Steiner, 1927 n. rank

Diagnosis: Draconematina. Nematodes usually less than 0.5 mm in length. Body arched dorsally and ventrally, usually sigmoid shaped. Greatest body width in anterior and posterior body regions. Cephalic sensory organs in 6 + (6 + 4) arrangement, paramphidial setae sometimes present. Inner circle of sensory organs obscure, papilloid; outer circles setiform. Lips obscure, no cuticularized supporting framework. Amphids spiral, usually located dorso-lateral. Cephalic adhesion tubes absent. Buccal cavity inconspicuous, rarely moderately developed or armed with dorsal tooth. Esophagus two part, corpus usually cylindrical; posterior bulb with well-developed valve. Cuticle with conspicuous, heavily cuticularized annules, except on rostrum and tail terminus, annules sometimes overlapping. Annules frequently ornamented with spines, ridges, grooves or subcuticular vacuoles. SS in

8 longitudinal rows. Stilt setae present or absent, if present in 4 or 6 subventral rows extending posteriorly from near mid-body. Stilt setae solid, non-glandular. Adhesion tubes absent. Tails of both sexes cylindrical-conoid.

One family: Epsilonematidae Steiner, 1927

Composition of the Suborder Draconematina de Coninck, 1965

SUPERFAMILY: Epsilonematoidea Steiner, 1927 n. rank
  FAMILY: Epsilonematidae Steiner, 1927
    SUBFAMILIES: Epsilonematinae Steiner, 1927
                       Glochinematinae Lorenzen, 1974

SUPERFAMILY: Draconematoidea Filipjev, 1918
  FAMILY: Draconematidae Filipjev, 1918
    Synonymy: Chaetosomatidae Schepotieff, 1907
               Drepanonematidae Johnston, 1938
               Claparediellidae Allgén, 1954
    SUBFAMILY: Draconematinae Filipjev, 1918
      Genus: *Draconema* Cobb, 1913 TYPE GENUS
        Synonymy: *Chaetosoma* Claparéde, 1863
                 Drepanonema Cobb, 1933
                 Claparediella Filipjev, 1934
    Nominal Species of the Genus *Draconema:*
      D. cephalatum Cobb, 1913 TYPE SPECIES
        Synonymy: *Chaetosoma cephalatum* (Cobb, 1913) Steiner, 1921
                 Tristicochaeta cephalatum (Cobb, 1913) Cobb, 1935
      D. haswelli (Irwin-Smith, 1918) Kreis, 1938
        Synonymy: *Chaetosoma haswelli* Irwin-Smith, 1918
                 Notochaetosoma haswelli (Irwin-Smith, 1918) Cobb, 1929
                 Drepanonema haswelli (Irwin-Smith, 1918) Cobb, 1933
                 Draconema bandaensis Kreis, 1938 n. syn.
                 Tristicochaeta haswelli (Irwin-Smith, 1918) Johnston, 1938
      D. claparedii (Mechnikov, 1867) Filipjev, 1918
        Synonymy: *Chaetosoma claparedii* Mechnikov, 1867
                 Chaetosoma armatum Giard and Barrois, 1874 n. syn.
                 Chaetosoma macrocephalum Schepotieff, 1907 n. syn.
                 Chaetosoma hibernicum Southern, 1914 n. syn.
                 Draconema armatum (Giard and Barrois, 1874) Filipjev, 1918 n. syn.
                 Draconema macrocephalum (Schepotieff, 1907) Filipjev, 1918 n. syn.
                 Notochaetosoma claparedii (Mechnikov, 1867) Cobb, 1929
                 Notochaetosoma armatum (Giard and Barrois, 1874) Cobb, 1929 n. syn.

Chaetosoma gronlandicum Levinsen, 1882 sp. inquir.

Synonymy: *Draconema gronlandicum* (Levinsen, 1882) Filipjev, 1918
  Notochaetosoma gronlandicum (Levinsen, 1882) Cobb, 1929
  Drepanonema gronlandicum (Levinsen, 1882) Cobb, 1933
  Claparediella gronlandicum (Levinsen, 1882) Filipjev, 1934

Discussion.—*Chaetosoma gronlandicum* was described by G. M. R. Levinsen in 1882 from Greenland. This species was described from adult nematodes, which were reported as having 3 longitudinal rows of posterior adhesion tubes. Since all adults in this superfamily have at least 4 rows, the 2 subventral rows were mistaken as being a single row. *Chaetosoma gronlandicum* resembles *Draconema ophicephalum* (Claparéde, 1863) Filipjev, 1918; but the evidence is so meager we have placed this species in *Species Inquirenda*.

Chaetosoma longuirostrum Schepotieff, 1907 sp. inquir.

Synonymy: *Chaetosoma longirostrum* Schepotieff, 1908
  Draconema longirostrum (Schepotieff, 1907) Filipjev, 1918
  Notochaetosoma longirostrum (Schepotieff, 1907) Cobb, 1929
  Drepanonema longirostrum (Schepotieff, 1907) Cobb, 1933
  Claparediella longirostrum (Schepotieff, 1907) Filipjev, 1934

Discussion.—In 1907, Schepotieff described *Chaetosoma longuirostrum,* and then in 1908 he redescribed this species changing the spelling of the species name to *C. longirostrum.* This species was collected from Naples and Bergen, Italy. He illustrates a second-stage larva, a male, and a female; and depicts 10 to 15 enlarged annules just posterior to the rostrum, which indicates this species is in the genus *Draconema*. The adults were reported as having 3 longitudinal rows of posterior adhesion tubes (adults of Draconematoidea have at least 4 rows), which means the 2 subventral rows were mistaken for 1 row. On the female, the 22 to 25 posterior adhesion tubes would approximate the number found on females of *Draconema ophicephalum* (Claparéde, 1863) Filipjev, 1918. The 25 posterior adhesion tubes on the male are greater than on any known species in the genus *Draconema*. On the basis of the description and illustrations, the number of adhesion tubes could involve a composite of the SlAT and SvAT. From the description and illustrations this species could not definitely be associated with any nominal species, and is here assigned to *Species Inquirenda*.

Chaetosoma meridionalis Kreis, 1938 sp. inquir.

Discussion.—H. A. Kreis in 1938 described *Chaetosoma meridionalis* from the south seas of the Pacific Ocean. From the inadequate description and illustrations we were unable to place this species, other than in the family Prochaetosomatidae; because of this the species is placed in *Species Inquirenda*.

KEY TO FAMILIES, SUBFAMILIES AND GENERA OF DRACONEMATOIDEA

1. Esophagus with swollen corpus and swollen posterior bulb (dumbbell-shaped). Buccal cavity obscure, collapsed, and unarmed . . . . Draconematidae Filipjev, 1918 *2*
  Corpus of esophagus cylindrical, sometimes with slight median swelling; esophagus with terminal

posterior swelling or bulb, with or without cuticularized valve. Buccal cavity collapsed to moderately developed, usually with dorsal tooth . . . . . Prochaetosomatidae n. fam. *5*

2. Males and females usually without cephalic sublateral acanthiform setae on rostrum. Females with or without sublateral posterior adhesion tubes (SlAT) on tail, if acanthiform setae on rostrum, then female tail always with SlAT. Males with or without large corniform setae anterior to subventral posterior adhesion tubes (SvAT) and preanally. *3*

Males and females with cephalic sublateral acanthiform setae on rostrum. Female tail without SlAT. Males with or without preanal acanthiform or corniform setae, but never located anterior to SvAT . . . . *Paradraconema* n. gen. (p. 43)

3. First body annules posterior to rostrum not differentiated from succeeding annules by size . . *4*

First 7 to 14 body annules posterior to rostrum much larger than succeeding annules . . . . *Draconema* Cobb, 1913 (p. 27)

4. Males and females with conspicuous, alternating, long and short, slender, adhesion tubes in rows of SlAT. Males with large corniform setae just anterior to SvAT and preanal . . . . *Dracotoranema* n. gen. (p. 89)

Males and females with long and short adhesion tubes not alternating in rows of SlAT. Males without corniform setae . . . . *Dracograllus* n. gen. (p. 63)

5. Buccal cavity collapsed or weakly developed, unarmed. Esophagus with inconspicuous terminal swelling. Cuticle very thick, thickness may obscure internal structures, rostral cuticle 4 to 12 μm thick . . . . Notochaetosomatinae n. subfam. *6*

Buccal cavity weak to moderately developed, armed with inconspicuous or conspicuous dorsal tooth. Esophagus with either conspicuous terminal swelling or posterior bulb with or without valve. Cuticle not unusually thick *7*

6. Cephalic adhesion tubes (CAT) on posterior half of and just posterior to rostrum; rostrum cuticle 4 to 5 μm thick . . . . *Notochaetosoma* Irwin-Smith, 1918 (p. 119)

Cephalic adhesion tubes on anterior half of rostrum; rostrum cuticle 7 to 12 μm thick . . . . *Dracogalerus* n. gen. (p. 124)

7. Amphids conspicuous, circular to loop-shaped. Buccal cavity collapsed or weak to moderately developed. Corpus of esophagus with either terminal swelling or posterior bulb with or without a valve *8*

Amphids inconspicuous, inverted "U" shape (staple-shaped), on base of rostrum and extending into first few body annules. Buccal cavity well developed, cylindrical. Corpus of esophagus with small median swelling. Posterior bulb with conspicuous cuticularized valve . . . Dracognominae n. subfam. (one genus *Dracognomus* n. gen. (p. 113)

8. Anterior 25 to 45% of body slender. Cephalic adhesion tubes on posterior part of rostrum or posterior to rostrum, with posterior row 1 or less rostral widths (4 to 11 annules) posterior to rostrum. Esophagus short, posterior bulb with or without cuticularized valve . . . . Prochaetosomatinae n. subfam. *9*

Anterior 50 to 59% of body slender. Cephalic adhesion tubes posterior to rostrum, with posterior row 2 to 3 rostral widths (26 to 37 annules) posterior to rostrum. Esophagus long, terminal swelling, without cuticularized valve . . . . Cygnonematinae n. subfam. (one genus *Cygnonema* n. gen. (p. 111)

9. Buccal cavity weakly developed (partially collapsed), with conspicuous or inconspicuous dorsal tooth, ventral teeth absent. Posterior esophageal bulb without cuticularized valve *10*

Buccal cavity moderately developed, with conspicuous dorsal tooth, with or without 1 to 2 obscure ventrally positioned teeth. Posterior esophageal bulb with cuticularized valve . . . . *Prochaetosoma* Micoletzky, 1922 (p. 91)

10. Rostrum conical. Esophageal region swollen dorsally and ventral side almost straight. Tail shape cylindrical-conoid, 7 to 13 tail annules . . . . *Draconactus* n. gen. (p. 105)

Rostrum broadly rounded. Esophageal region without dorsal swelling, nearly cylindrical. Tail shape elongate cylindrical-conoid (spike-like), 3 to 4 tail annules . . . . *Apenodraconema* n. gen. (p. 108)

Notochaetosoma macrocephalum (Schepotieff, 1907) Cobb, 1929 n. syn.
Notochaetosoma hibernicum (Southern, 1914) Cobb, 1929 n. syn.
Drepanonema claparedii (Mechnikov, 1867) Cobb, 1933
Drepanonema armatum (Giard and Barrois, 1874) Cobb, 1933 n. syn.
Drepanonema macrocephalum (Schepotieff, 1907) Cobb, 1933 n. syn.
Drepanonema hibernicum (Southern, 1914) Cobb, 1933 n. syn.
Claparediella claparedii (Mechnikov, 1867) Filipjev, 1934
Claparediella macrocephala (Schepotieff, 1907) Filipjev, 1934 n. syn.
Draconema macrocephalum (Schepotieff, 1908) Schuurmans Stekhoven, 1935 n. syn.
Draconema hibernicum (Southern, 1914) Kreis, 1938 n. syn.

D. ophicephalum (Claparéde, 1863) Filipjev, 1918
  Synonymy: *Chaetosoma ophicephalum* Claparéde, 1863
  Chaetosoma annulatum Ditlevsen, 1915 n. syn.
  Draconema cephalatum Steiner, 1916, *nec.* Cobb, 1913 n. syn.
  Draconema cephalatum Filipjev, 1918, *nec.* Cobb, 1913 n. syn.
  Draconema ponticum Filipjev, 1918 n. syn.
  Draconema micoletzkyi Kreis, 1928 n. syn.
  Notochaetosoma ophicephalum (Claparéde, 1863) Cobb, 1929
  Notochaetosoma annulatum (Ditlevsen, 1915) Cobb, 1929 n. syn.
  Drepanonema ophicephalum (Claparéde, 1863) Cobb, 1933
  Drepanonema annulatum (Ditlevsen, 1915) Cobb, 1933 n. syn.
  Claparediella ophicephala (Claparéde, 1863) Filipjev, 1934
  Claparediella annulatum (Ditlevsen, 1915) Filipjev, 1934 n. syn.
  Draconema annulatum (Ditlevsen, 1915) Schuurmans Stekhoven, 1935 n. syn.
  Tristicochaeta ponticum (Filipjev, 1918) Cobb, 1935 n. syn.
  Tristicochaeta micoletzkyi (Kreis, 1928) Cobb, 1935 n. syn.

D. antarcticum n. sp.
D. chilense n. sp.

Genus: *Paradraconema* n. gen.

Nominal Species of the Genus *Paradraconema:*
  P. floridense n. sp. TYPE SPECIES
  P. spinosum (Southern, 1914) n. comb.
    Synonymy: *Chaetosoma spinosum* Southern, 1914 n. syn.
      Notochaetosoma spinosum (Southern, 1914) Cobb, 1929 n. syn.
      Drepanonema spinosum (Southern, 1914) Cobb, 1933 n. syn.
      Claparediella spinosum (Southern, 1914) Filipjev, 1934 n. syn.
      Draconema spinosum (Southern, 1914) Kreis, 1938 n. syn.
  P. californicum n. sp.
  P. newelli n. sp.
  P. singaporense n. sp.
  P. meridionale (Kreis, 1938) n. comb.
    Synonymy: *Draconema meridionalis* Kreis, 1938 n. syn.
  P. hopperi n. sp.
  P. antarcticum n. sp.
Genus: *Dracograllus* n. gen.
Nominal Species of the Genus *Dracograllus:*
  D. cobbi n. sp. TYPE SPECIES
  D. mawsoni n. sp.
  D. demani n. sp.
  D. filipjevi n. sp.
  D. timmi n. sp.
  D. chitwoodi n. sp.
  D. kreisi n. sp.
  D. gerlachi n. sp.
  D. stekhoveni n. sp.
  D. wieseri n. sp.
  D. eira (Inglis, 1968) n. comb.
    Synonymy: *Draconema eira* Inglis, 1968 n. syn.
  D. falcatus (Irwin-Smith, 1918) n. comb.
    Synonymy: *Chaetosoma falcatum* Irwin-Smith, 1918 n. syn.
      Notochaetosoma falcatum (Irwin-Smith, 1918) Cobb, 1929 n. syn.
      Drepanonema falcatum (Irwin-Smith, 1918) Cobb, 1933 n. syn.
      Claparediella falcatum (Irwin-Smith, 1918) Filipjev, 1934 n. syn.
      Draconema falcatum (Irwin-Smith, 1918) Kreis, 1938 n. syn.
      Tristicochaeta falcata (Irwin-Smith, 1918) Johnston, 1938 n. syn.
  D. solidus (Gerlach, 1952) n. comb.
    Synonymy: *Draconema solidum* Gerlach, 1952 n. syn.

Genus: *Dracotoranema* n. gen.
Nominal Species of the Genus *Dracotoranema:*
   D. trispinosum n. sp. TYPE SPECIES

FAMILY: Prochaetosomatidae n. fam.
  SUBFAMILY: Prochaetosomatinae n. subfam.
    Genus: *Prochaetosoma* Micoletzky, 1922 TYPE GENUS
    Nominal Species of the Genus *Prochaetosoma:*
     P. primitiva (Steiner, 1916) Micoletzky, 1922 TYPE SPECIES
       Synonymy: *Chaetosoma primitivum* Steiner, 1916
              Draconema primitivum (Steiner, 1916) Filipjev, 1918
              Drepanonema primitivum (Steiner, 1916) Cobb, 1933 n. syn.
              Claparediella primitivum (Steiner, 1916) Filipjev, 1934 n. syn.
     P. cayense n. sp.
     P. vitielloi n. sp.
     P. mediterranicum n. sp.
     P. arcticum (Kreis, 1963) n. comb.
       Synonymy: *Draconema arcticum* Kreis, 1963 n. syn.
     P. campbelli (Allgén, 1932) n. comb.
       Synonymy: *Chaetosoma campbelli* Allgén, 1932 n. syn.
              Drepanonema campbelli (Allgén, 1932) Cobb, 1933 n. syn.
              Claparediella campbelli (Allgén, 1932) Filipjev, 1934 n. syn.
              Draconema campbelli (Allgén, 1932) Schuurmans Stekhoven, 1935 n. syn.
     P. longicapitata (Allgén, 1932) n. comb.
       Synonymy: *Chaetosoma longicapitata* Allgén, 1932 n. syn.
              Drepanonema longicapitata (Allgén, 1932) Cobb, 1933 n. syn.
              Claparediella longicapitata (Allgén, 1932) Filipjev, 1934 n. syn.
              Draconema longicapitata (Allgén, 1932) Schuurmans Stekhoven, 1935 n. syn.
     P. lugubre (Gerlach, 1957) n. comb.
       Synonymy: *Drepanonema lugubre* Gerlach, 1957 n. syn.
    Genus: *Draconactus* n. gen.
    Nominal Species of the Genus *Draconactus:*
     D. cutus n. sp. TYPE SPECIES
     D. suillus (Allgén, 1932) n. comb.
       Synonymy: *Chaetosoma suilla* Allgén, 1932 n. syn.
              Drepanonema suilla (Allgén, 1932) Cobb, 1933 n. syn.
              Claparediella suilla (Allgén, 1932) Filipjev, 1934 n. syn.
              Draconema suilla (Allgén, 1932) Schuurmans Stekhoven, 1935 n. syn.

Genus: *Apenodraconema* n. gen.
Nominal Species of the Genus *Apenodraconema*:
  A. chlidosis n. sp. TYPE SPECIES
  A. spinicaudum (Gerlach, 1958) n. comb.
    Synonymy: *Draconema spinicauda* Gerlach, 1958 n. syn.
SUBFAMILY: Cygnonematinae n. subfam.
  Genus: *Cygnonema* n. gen. TYPE GENUS
  Nominal Species of the Genus *Cygnonema*:
    C. steineri n. sp. TYPE SPECIES
SUBFAMILY: Dracognominae n. subfam.
  Genus: *Dracognomus* n. gen. TYPE GENUS
  Nominal Species of the Genus *Dracognomus*:
    D. marioni n. sp. TYPE SPECIES
    D. simplex (Gerlach, 1954) n. comb.
      Synonymy: *Draconema simplex* Gerlach, 1954 n. syn.
    D. notohalensis n. sp.
SUBFAMILY: Notochaetosomatinae n. subfam.
  Genus: *Notochaetosoma* Irwin-Smith, 1918 TYPE GENUS
  Nominal Species of the Genus *Notochaetosoma*:
    N. tenax Irwin-Smith, 1918 TYPE SPECIES
  Genus: *Dracogalerus* n. gen.
  Nominal Species of the Genus *Dracogalerus*:
    D. afrikaanus n. sp. TYPE SPECIES
    D. bastiani n. sp.
    D. cryptocephalus (Irwin-Smith, 1918) n. comb.
      Synonymy: *Notochaetosoma cryptocephalum* Irwin-Smith, 1918 n. syn.

Species Inquirendae of the Superfamily Draconematoidea

Tristicochaeta inarimense Panceri, 1878 sp. inquir.

Synonymy: *Chaetosoma tristicochaeta* (Panceri, 1878) Schepotieff, 1908
Drepanonema tristicochaeta (Panceri, 1878) Cobb, 1933
Claparediella tristicochaeta (Panceri, 1878) Filipjev, 1934
Draconema tristicochaeta (Panceri, 1878) King, 1939
Chaetosoma inarimense (Panceri, 1878) Allgén, 1939

Discussion.—Because of the inadequate descriptions and illustrations, we have concluded that this species is unidentifiable and is a *Species Inquirenda* (for further details concerning this genus and species see Taxonomic History).

In 1907 and 1908 Schepotieff described *Chaetosoma macrocephalum* and *C. longirostrum*. He wrote at length on the morphology of the species he had studied and concluded that *Tristicochaeta* Panceri, 1876 was congeneric with *Chaetosoma* Claparéde, 1863. As a result of this conclusion he proposed the new name *Chaetosoma tristicochaeta* for the nematode described by Panceri in 1876 as *T. inarimense*. This was an invalid emendation of this species name by Schepotieff; also he was not aware that the generic name *Chaetosoma* Claparéde, 1863 was preoccupied. In 1939, Allgén corrected this error by making *T. inarimense* Panceri, 1876 a synonym of *Chaetosoma inarimense* (Panceri, 1876).

Genus *Draconema* Cobb, 1913

Syn: *Chaetosoma* Claparéde, 1863
Drepanonema Cobb, 1933
Claparediella Filipjev, 1934

Diagnosis Emended: Draconematinae. Nematodes 0.6 to 2.2 mm long. Swollen esophageal region averaging 12% total body length. Females greatest body width at mid-body, males usually at esophageal region. Rostrum ornamented with subcuticular markings, setae present. Amphids large, conspicuous, elongate loop-shaped to unispiral, slightly dorsal on rostrum. Female amphids dorsal arm usually longer than ventral arm. Male amphids ventral arm usually longer than dorsal arm. Twelve CAT, paired, in 2 transverse rows on rostrum. First 7 to 14 body annules posterior to rostrum larger than succeeding annules, with anteriorly directed margins, ornamented with conspicuous subcuticular markings and vacuoles. Remaining annulation marked with subcuticular granulation. Two pairs of paravulval setae, 1 pair anterior and 1 posterior to vulva (figs. 22–23). With or without spine-like projections surrounding vulva (fig. 22). Males with 4 pairs of anal setae, usually 2 pairs anterior and 2 posterior to anus. Anal setae generally uniformly tapered (fig. 24); but some or all setae can be short, with round broad base, and unevenly tapered (figs. 5, 48). All PAT anterior to anus, paired or unpaired, numbers of tubes variable between species, setae intermingled with tubes. Anal flap may be faintly crenated, crenations not as pronounced as in *Paradraconema* n. gen. Setae on non-annulated tail region variable in number and location.

Type Species: *Draconema cephalatum* Cobb, 1913

Differs from other genera in Draconematinae by 7 to 14 enlarged annules just posterior to rostrum. Differs from *Paradraconema* n. gen. by absence of eyespots, cephalic sublateral acanthiform setae on rostrum, and preanal acanthiform or corniform setae in males; from *Dracograllus* n. gen. and *Dracotoranema* n. gen. by absence of SlAT on tail. Differs from *Dracograllus* n. gen. by absence of any SS borne on conspicuous, raised, cuticular pedicels. Differs from *Dracotoranema* n. gen. by smaller amphids; absence of conspicuous alternating long and short, slender, adhesion tubes in rows of SlAT; and absence of preanal corniform setae and ventral corniform seta anterior to SvAT in males.

Larval Stages: (figs. 33–41). All 4 larval stages available in *Draconema*. First-stage larvae 0.3 to 0.4 mm long. Similar to adults. Relaxed body nearly straight or curved ventrally. Esophageal region swollen, rest of body nearly cylindrical. Lip region with 6 modified setae, 2 subdorsal, 2 subventral and 2 lateral; setae broad-based, with distal ends directed ventrally, nearly triangular in shape. Four cephalic setae, 2 subdorsal and 2 subventral, just posterior to triangular setae on lip region. Rostrum absent. Amphids small, circular, slightly dorsal. CAT absent. Esophagus with swollen corpus and posterior bulb, isthmus not as distinct as in other stages. Entire body length with fine annulation, except for non-annulated tail region. Most specimens without SS; some with 2 lateral pairs of long setae, 1 pair midway on swollen esophageal region and 1 pair either just anterior to anus or midway on tail. Posterior adhesion tubes absent. Caudal glands and spinneret not observed. Anal flap absent. Non-annulated tail region without punctuations and setae. Differentiated from other 3 stages by absence of cephalic and posterior adhesion tubes, unable to differentiate between species.

Second-stage larvae 0.3 to 0.6 mm long. Similar to adults. Relaxed body either nearly straight or arched ventrally and dorsally. Esophageal region swollen, rest of body nearly cylindrical. Rostrum with very fine indistinct ornamentation, setae present. Amphids large, conspicuous, lateral on rostrum; elongate loop-shaped to unispiral, sometimes ventral arm starts to form double spiral. Dorsal arm of amphid usually longer than ventral arm. Single dorsal CAT on rostrum. One to 4 indistinct large annules just posterior to rostrum, margins anteriorly directed, annules without ornamentation. Six rows of SS on swollen esophageal region, 4 sublateral, 1 ventral and 1 dorsal; setae in 4 sublateral rows on remainder of body. Two pairs of SlAT in 2 longitudinal rows. Caudal glands and spinneret usually indistinct. Anal flap present. Non-annulated tail region finely ornamented with punctations, setae variable in number and location. Differentiated from other 3 stages by 1 CAT, 2 pairs of SlAT in 2 rows.

Third-stage larvae 0.4 to 0.8 mm long. Similar to adults. Esophageal region swollen, rest of body nearly cylindrical. Amphids large, conspicuous, lateral on rostrum; elongate loop-shaped to unispiral, most tend toward unispiral. Dorsal arm of amphid usually longer than ventral arm. Three CAT, 2 subdorsal and 1 dorsal, in 2 transverse rows on rostrum. Rostrum with setae. One to 4 large annules just posterior to rostrum, ornamentation varies from none to very fine, margins anteriorly directed. Eight rows of SS on swollen esophageal region, 4 sublateral, 2 subdorsal and 2 subventral. Five rows of SS on mid-body region, 4 sublateral and 1 dorsal; 4 sublateral rows on annulated tail region. Some rows with alternating long and short setae. Three pairs of SlAT in 2 longitudinal rows. Anal flap present. Setae on non-annulated tail region variable in number and location. Differentiated from other 3 stages by 3 CAT, 3 pairs of SlAT in 2 rows.

Fourth-stage larvae 0.5 to 1.4 mm long. Similar to adults. Esophageal region swollen, young females swollen at mid-body. Amphids large, conspicuous, lateral on rostrum, elongate loop-shaped or unispiral. Dorsal arm of amphid usually longer than ventral arm. Four CAT, 2 sublateral and 2 subdorsal, in 2 transverse rows on rostrum. Rostrum with setae. Two to 7 large annules just posterior to rostrum, ornamentation not as distinct as in adults, margins directed anteriorly. Eight rows of SS on swollen esophageal region, 4 sublateral, 2 subdorsal and 2 subventral. Seven rows of SS on mid-body region, 4 sublateral, 2 subdorsal, 1 ventral; 4 sublateral rows on annulated tail region. Some rows with alternating long and short setae. Three longitudinal rows of posterior adhesion tubes, 2 sublateral and 1 ventral. Five pairs of SlAT, VAT number variable between species. Anal flap sometimes slightly crenate. Setae on non-annulated tail region variable in number and location. Differentiated from other 3 stages by 4 CAT, 3 rows of PAT.

Discussion.—(For details of the status of the generic name *Draconema* Cobb, 1913, see Taxonomic History.)

Key to Species of *Draconema* Cobb, 1913

1. Males with 14 to 19 SlAT.[4] Females with 18 to 24 SlAT *2*
   Males with 7 to 10 SlAT. Females with 9 to 16 SlAT *3*
2. Male anal setae short, prominent, with round broad base, and with uneven taper (fig. 48). Females 0.6 to 1.1 mm long, without conspicuous spine-like projections surrounding vulva . . . . *chilense* n. sp. (p. 40)
   Male anal setae uniformly tapered (fig. 24). Females 1.2 to 1.7 mm long, with conspicuous spine-like projections surrounding vulva . . . *ophicephalum* (Claparéde, 1863) Filipjev, 1918 (p. 36)
3. Males and females with 1 long subdorsal or subventral pair of setae on non-annulated tail region (other setae shorter), long setae equal in length to long setae on annulated tail region, if subdorsal usually just posterior to last complete tail annule. Males either with uniformly tapered anal setae; or with or without some short, broad-based anal setae with uneven taper. Females with 11 to 16 SlAT, amphids elongate loop-shaped or unispiral, paravulval setae 11 to 14 μm long . . *4*
   Males and females without long setae on non-annulated tail region, setae not equal in length to long setae on annulated tail region. Male anal setae uniformly tapered. Females with 9 to 11 SlAT; amphids usually unispiral, sometimes ventral arm only partially converged on dorsal arm; paravulval setae 8 to 9 μm long . . . . . *claparedii* (Mechnikov, 1867) Filipjev, 1918 (p. 33)
4. Males with 1 long subdorsal pair of setae just posterior to last complete tail annule, equal in length to long setae on annulated tail region; first SlAT 41 to 65 μm long; first SvAT 34 to 58 μm long. Females with 11 to 14 SlAT, amphids elongate loop-shaped or unispiral *5*
   Males with 1 long subventral pair of setae about 50% on non-annulated tail region, equal in length to long setae on annulated tail; first SlAT 70 to 86 μm long; first SvAT 62 to 80 μm long. Females with 14 to 16 SlAT, amphids elongate loop-shaped, never unispiral . . . . *antarcticum* n. sp. (p. 39)
5. Males with 2 pairs of uniformly tapered anal setae; and 2 pairs of short, broad-based, unevenly tapered anal setae. Females with 5 pairs of setae on non-annulated tail region; amphids usually

---

4. All SlAT counts and measurements on right side of body.

unispiral, if loop-shaped ventral arm always partially converged toward dorsal arm . . . .
. . . . *cephalatum* Cobb, 1913 (p. 29)

Male anal setae all uniformly tapered. Females with 4 pairs of setae on non-annulated tail region; amphids usually elongate loop-shaped with distal ends of arms open, sometimes ventral arm converging toward dorsal arm almost forming unispiral . . . .
. . . . *haswelli* (Irwin-Smith, 1918) Kreis, 1938 (p. 30)

Draconema cephalatum Cobb, 1913

(Figs. 1 to 8, 198 to 199)

Syn: *Chaetosoma cephalatum* (Cobb, 1913) Steiner, 1921
  Tristicochaeta cephalatum (Cobb, 1913) Cobb, 1935

Measurements (25 ♀♀): L = 1.1 (0.9–1.4) mm; b = 10.4 (8.3–14.0); c = 13.8 (8.7–18.0); V = 54 (50–60)%; CAT = 25 (20–33) μm; SER (L/W) = 1.9 (1.5–2.2); SER (E/L) = 11 (9–14)%; No Large SER Ann = 11 (9–13); SS = 14–73 μm; first SlAT = 49 (41–63) μm; last SlAT = 40 (33–54) μm; first SvAT = 43 (33–59) μm; last SvAT = 29 (23–39) μm; No SlAT = 14 (12–14); No SvAT = 16 (14–18); Non-ann Term to Tail Length = 51 (44–57)%; T/ABD = 4.0 (3.1–4.8).

(20 ♂♂): L = 1.1 (0.8–1.5) mm; a = 19.4 (16.3–24.0); b = 11.2 (7.3–13.8); c = 10.8 (8.0–12.5); CAT = 25 (21–32) μm; SER (L/W) = 1.9 (1.7–2.3); SER (E/L) = 11 (9–14)%; No Large SER Ann = 11 (9–13); SS = 15–82 μm; first SlAT = 51 (41–64) μm; last SlAT = 42 (35–51) μm; first SvAT = 43 (34–53) μm; last SvAT = 30 (22–38) μm; No SlAT = 9 (8–10); No SvAT = 16 (14–18); Non-ann Term to Tail Length = 37 (31–45)%; T/ABD = 4.0 (3.5–5.0); Spic = 72 (64–85) μm; Gub = 20 (16–23) μm.

Males Emended.—Amphids elongate loop-shaped. CAT typical of genus. Longest SS on swollen esophageal region. Long and short SS intermingled with SlAT. Three to 6 long setae with SlAT, usually 4, almost equal in length to adjacent tube, not alternating with tubes. Caudal glands extend anterior to anus 2.7 to 2.9 times ABD. Anal setae with 2 subventral pairs, short, broad-based, unevenly tapered, 6 to 9 μm long, 1 pair anterior and 1 posterior to anus; and 2 uniformly tapered sublateral pairs, 8 to 13 μm long, 1 pair anterior and 1 posterior to anus (fig. 5). Anal flap short, crenate. Six pairs of setae on non-annulated tail region, position measured from last complete tail annule to tail tip; 1 subventral pair about 50%; 3 subdorsal pairs, 1 long pair just posterior to last complete tail annule equal in length to long setae on annulated tail, 1 short pair just posterior to long pair, and 1 short pair about 33%; 2 lateral pairs, 1 about 66%, 1 about 75%.

Females Emended.—Similar to males. Amphids usually unispiral, or if elongate loop-shaped sometimes ventral arm converging toward dorsal arm. Vulva encircled by minute spine-like projections, smaller but similar to *D. ophicephalum*. Paravulval setae 11 to 13 μm long. Longest SS on swollen esophageal region. Short setae intermingled with SlAT. Caudal glands extend anterior to anus 2.3 to 4.9 times ABD. Setae on non-annulated tail region as in males, except with only 1 lateral pair about 50%.

First-Stage Larvae.—Not observed.

Second-Stage Larvae.—L = 0.4 mm. Similar to adults. Amphids unispiral. Two large annules posterior to rostrum. CAT and SlAT typical of genus. Two pairs of setae on non-annulated tail region; 1 lateral pair about 50%, 1 pair adjacent to or slightly dorsal to lateral pair.

Third-Stage Larvae.—L = 0.4 to 0.6 mm. Similar to adults. Amphids unispiral. One to 2 large annules posterior to rostrum. CAT and SlAT typical of genus. Three to 4 pairs of setae on non-annulated tail region; 1 lateral pair about 50%; 1 subventral pair present or absent adjacent to long subdorsal pair; 2 subdorsal pairs, 1 long pair just posterior to last complete tail annule equal in length to long setae on annulated tail, 1 short pair near tail tip.

Fourth-Stage Larvae.—L = 0.8 to 0.9 mm. Similar to adults. Amphids unispiral. Two to 4 large annules posterior to rostrum. CAT and SlAT typical of genus; 9 VAT. Four pairs of setae on non-annulated tail region; 1 lateral pair about 66%; 3 subdorsal pairs, 1 long pair just posterior to last complete tail annule equal in length to long setae on annulated tail, 1 short pair just posterior to long pair, 1 short pair about 33%.

Lectotype (♂).—Collected in 1910 by Dr. N. A. Cobb. Catalogue No. T-241t, USDA Nematode Collection, Beltsville, Maryland.

Paralectotype.—1 ♂. Same data as lectotype, USDA Nematode Collection, Beltsville, Maryland.

Type Habitat.—Marine, associated with sand at base of algae.

Type Locality.—Kingston Harbor and an Island off Port Royal, Jamaica, West Indies.

Distribution.—Known distribution. Antarctica. Australia: Christies Beach, Norman Ville Beach, Port Jackson, and Port Noarlunga. Solano, Colombia. Punta Carnera, Ecuador. Florida, USA: Bear Cut, Coral Key, and Soldier Key. Japan: Akkeshi, Hokkaida; Ibusuki, Kyushu; Nakaminato, Ibaraki-ken; Shimoda, Shizuoka-ken; and Wajima, Ishikawa-ken. Tamuning, Agana Bay, Guam Island, Marianas Islands. Galeta Beach and Panama City, Panama. Philippine Islands: Little Santa Cruz Island; Matabungkay, Batangas; and Zamboanga, Mindanao. Society Islands: Tiarei, Papeete, Tahiti Island; Uturoa and Raiatea, Raiatea Island. Kingston Harbor and an island off Port Royal, Jamaica, West Indies.

This species has also been recorded from Fuegian Archipelago, Antarctica; Barents Sea; Campbell Island; Hudson Bay, Canada; Falkland Islands; Kerguelen Island; North Sea; Norway; Öresund; Red Sea; Skagerrak; South Georgia Islands; Svalbard; California, USA; Tobago, West Indies.

Diagnosis.—Males differ from other known *Draconema* males by 2 pairs of uniformly tapered anal setae; and 2 pairs of short, broad-based, unevenly tapered anal setae. Females most closely resemble *D. haswelli* but differ by 5 pairs of setae on non-annulated tail region; amphids usually unispiral, or if loop-shaped, ventral arm always partially converged toward dorsal arm. Females differ from *D. chilense* n. sp., *D. ophicephalum*, and *D. antarcticum* n. sp. by fewer SlAT; and from *D. claparedii* by 1 long pair of setae on non-annulated tail region.

Second-stage larvae differ from other known second-stage *Draconema* by 2 pairs of short setae on non-annulated tail region. Third-stage larvae differ from other known third-stage *Draconema* by 1 pair of short setae on non-annulated tail region near tail tip. Fourth-stage larvae differ from other known fourth-stage *Draconema* by 1 short subdorsal pair of setae about 33%, and 1 short lateral pair of setae about 66% on non-annulated tail region.

The syntype series of N. A. Cobb's was available for study. These specimens were in fair condition and conform to Cobb's original description, therefore, the lectotype is not redescribed in this review.

Draconema haswelli (Irwin-Smith, 1918) Kreis, 1938

(Figs. 10, 12, 15 to 16, 18)

Syn: *Chaetosoma haswelli* Irwin-Smith, 1918
Notochaetosoma haswelli (Irwin-Smith, 1918) Cobb, 1929
Drepanonema haswelli (Irwin-Smith, 1918) Cobb, 1933 n. syn.
Tristicochaeta haswelli (Irwin-Smith, 1918) Johnston, 1938
Draconema bandaensis Kreis, 1938 n. syn.

Measurements (20 ♀♀): L = 1.4 (1.0–1.7) mm; b = 11.4 (8.5–14.2); c = 13.7 (10.2–17.3); V = 53 (47–56)%; CAT = 29 (26–32) μm; SER (L/W) = 2.3 (1.9–2.6); SER (E/L) = 11 (8–14)%; No Large SER Ann = 11 (9–13); SS = 16–72 μm; first SlAT = 58 (54–65) μm; last SlAT = 44 (40–48) μm; first SvAT = 51 (46–55) μm; last SvAT = 32 (28–36) μm; No SlAT = 12 (11–13); No SvAT = 17 (15–19); Non-ann Term to Tail Length = 51 (42–60)%; T/ABD = 4.6 (3.6–5.7).

(20 ♂♂): L = 1.3 (1.0–1.6) mm; a = 31.2 (23.1–36.4); b = 11.8 (9.0–14.0); c = 10.2 (8.0–14.0); CAT = 30 (26–34) μm; SER (L/W) = 2.3 (2.0–2.6); SER (E/L) = 10 (9–13)%; No Large SER Ann = 10 (9–13); SS = 15–61 μm; first SlAT = 61 (58–65) μm; last SlAT = 46 (41–50) μm; first SvAT = 52 (49–58) μm; last SvAT = 32 (28–36) μm; No SlAT = 8 (7–10); No SvAT = 16 (14–18); Non-ann Term to Tail Length = 33 (25–40)%; T/ABD = 4.8 (4.0–5.5); Spic = 70 (62–75) μm; Gub = 19 (17–21) μm.

Males Emended.—Amphids elongate loop-shaped. CAT typical of genus. Longest SS on swollen esophageal region. Long and short setae intermingled with SlAT. Three to 4 long setae with SlAT, usually 4, longer than adjacent tubes, not alternating with tubes. Caudal glands extend anterior to anus 3.3 to 4.6 times ABD. Anal setae evenly tapered, sublateral, 17 to 26 μm long, 2 pairs anterior and 2 posterior to anus (fig. 12). Anal flap short, slightly crenated. Five pairs of setae on non-annulated tail region, position measured from last complete tail annule to tail tip; 1 subventral pair about 50%; 1 lateral pair adjacent to subventral pair; 3 subdorsal pairs, 1 long pair just posterior to last complete tail annule equal in length to long setae on annulated tail region, 1 short pair just posterior to long subdorsal pair, 1 short pair about 33%.

Females Emended. — Similar to males. Amphids usually elongate loop-shaped with distal ends of arms open, sometimes ventral arm converging toward dorsal arm almost forming unispiral. Vulva encircled with minute spine-like projections, smaller but similar to *D. ophicephalum*. Paravulval setae 12 to 14 μm long. Longest SS on swollen esophageal region. Short setae intermingled with SIAT. Caudal glands extend anterior to anus 3.8 to 5.2 times ABD. Anal flap long, crenate. Four pairs of setae on non-annulated tail region; 1 lateral pair about 66%; 3 subdorsal pairs, 1 long pair just posterior to last complete tail annule equal in length to long setae on annulated tail region, 1 short pair about 25%, 1 short pair about 50%.

First-Stage Larvae. — L = 0.3 mm. (See Generic Description.)

Second-Stage Larvae. — L = 0.3 to 0.4 mm. Similar to adults. Amphids unispiral. Two to 3 large annules posterior to rostrum. CAT and SlAT typical of genus. A single short, dorsal seta on non-annulated tail region, about 50%.

Third-Stage Larvae. — L = 0.6 mm. Similar to adults. Amphids unispiral. Two to 3 large annules posterior to rostrum. CAT and SlAT typical of genus. Three pairs of setae on non-annulated tail region; 1 lateral pair about 25%; 1 long subdorsal pair just posterior to last complete tail annule equal in length to long setae on annulated tail region; 1 short subdorsal pair about 25%.

Fourth-Stage Larvae. — L = 0.9 to 1.0 mm. Similar to adults. Two to 4 large annules just posterior to rostrum. CAT and SlAT typical of genus; 9 VAT. Three to 4 pairs of setae on non-annulated tail region; 1 lateral pair about 50%; 2 to 3 subdorsal pairs, 1 long pair just posterior to last complete tail annule equal in length to long setae on annulated tail region, 1 short pair just posterior to long subdorsal pair, 1 short pair present or absent about 75%.

Type Specimens (1 ♀, 1 ♂) — Collected in 1916 by V. A. Irwin-Smith. Catalogue Nos. W454 and W455, Australian Museum, Sydney, New South Wales, Australia.

Type Habitat. — Marine, associated with bottom debris, just below tidal zone.

Type Locality. — Port Jackson, N.S.W., Australia.

Distribution. — Known distribution. Port Jackson, Australia.

This species has also been recorded from Broken Bay, Australia; Red Sea; and the South Pacific Ocean (area of Indonesia).

Diagnosis. — Most closely resembles *D. cephalatum* differs by all uniformly tapered anal setae in males; females by 4 pairs of setae on non-annulated tail region, and amphids usually elongate loop-shaped with distal ends of arms open, sometimes arms partially converged almost forming unispiral. Differs from *D. chilense* n. sp. and *D. ophicephalum* by fewer SlAT, from *D. claparedii* by 1 long pair of setae on non-annulated tail region. Differs from *D. antarcticum* n. sp. by 1 long subdorsal pair of setae on non-annulated tail region in males; females differ by fewer SlAT, and 4 pairs of setae on non-annulated tail region.

Second-stage larvae differ from other known second-stage *Draconema* by single dorsal seta about 50% on non-annulated tail region. Third-stage larvae differ from other known third-stage *Draconema* by 1 lateral pair of setae about 25% and 1 subdorsal pair of setae about 66% on non-annulated tail region. Fourth-stage larvae differ from fourth stages of *D. cephalatum*, *D. claparedii* and *D. ophicephalum* by elongate loop-shaped amphid; from *D. antarcticum* n. sp. by smaller size; and from *D. chilense* n. sp. by larger size.

Type specimens were available for study, but these were in very poor condition. By comparing specimens collected from the type locality and the type specimens, it was possible to distinguish *D. haswelli* from other known species of *Draconema*.

Discussion. — *Draconema bandaensis* was described by H. A. Kreis in 1938, from the south seas of the Pacific Ocean. In the description he compared *D. bandaensis* with *D. cephalatum* Cobb, 1913, and concluded that they were different species. However, the original description and illustrations of *D. bandaensis* indicate the specimens were really *D. haswelli*.

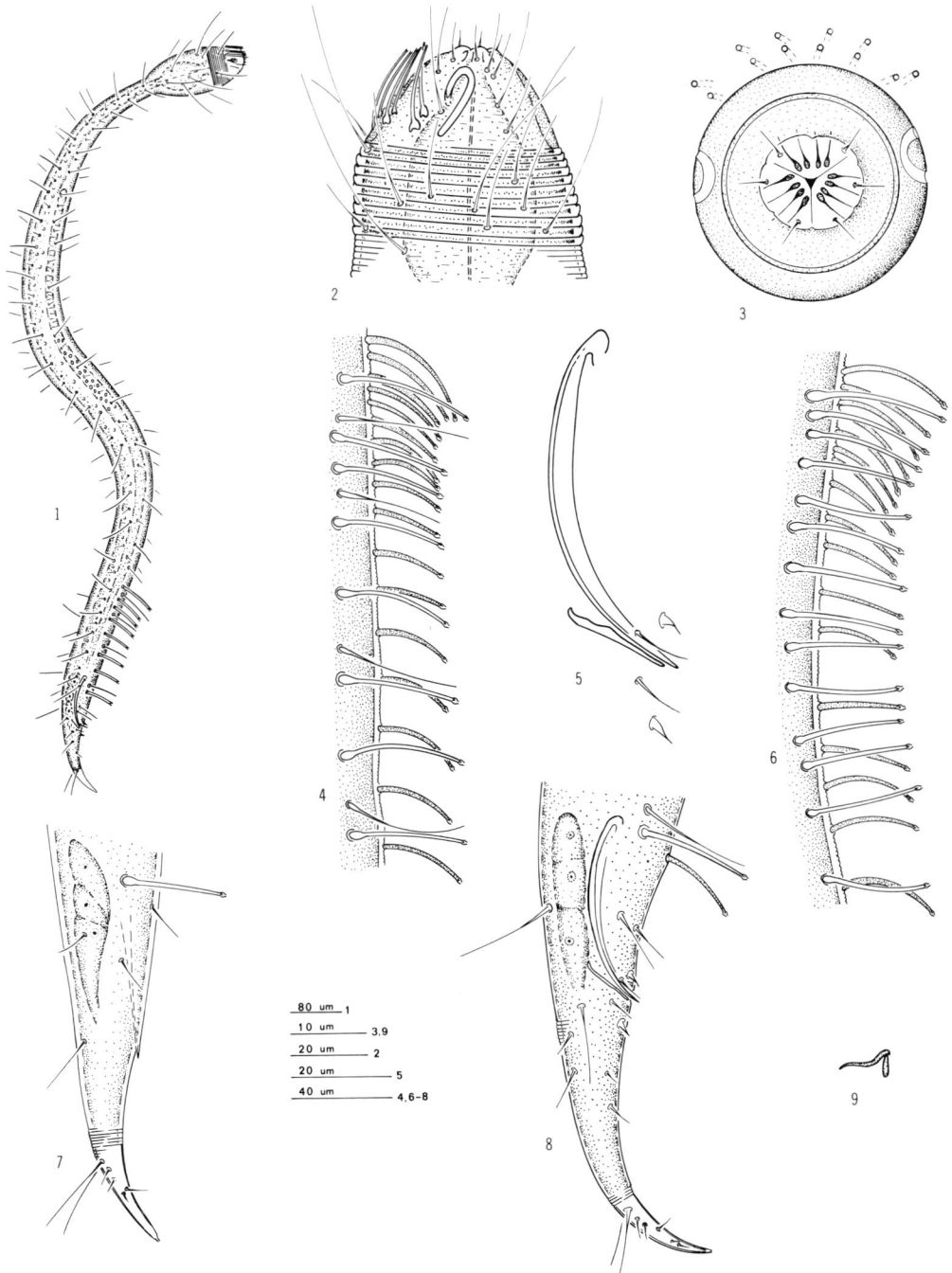

1. Figs. 1-8: *Draconema cephalatum* Cobb. 1) Male, full length[a], SvAT and correct number of CAT not illustrated; 2) Male, head; 3) Female, face view; 4) Male, PAT[b]; 5) Male, anal region; 6) Female, PAT[b]; 7) Female, tail; 8) Male, tail. Fig. 9: *Draconema antarcticum* n. sp., male, lateral view of internal projecting cuticularized rib of lip region.

a. Total number and position of setae on non-annulated tail region not illustrated.
b. Short SS with SlAT not illustrated.

During this same year (1938) T. H. Johnston, in the same publication in which he proposed the new family Drepanonematidae, made the new combination *Tristicochaeta haswelli* (Irwin-Smith, 1918). (For further details, see Taxonomic History.)

Draconema claparedii (Mechnikov, 1867) Filipjev, 1918

(Figs. 11, 13 to 14, 17, 19)

Syn: *Chaetosoma claparedii* Mechnikov, 1867
    Chaetosoma armatum Giard and Barrois, 1874 n. syn.
    Chaetosoma macrocephalum Schepotieff, 1907 n. syn.
    Chaetosoma hibernicum Southern, 1914 n. syn.
    Draconema armatum (Giard and Barrois, 1874) Filipjev, 1918 n. syn.
    Draconema macrocephalum (Schepotieff, 1907) Filipjev, 1918 n. syn.
    Notochaetosoma claparedii (Mechnikov, 1867) Cobb, 1929
    Notochaetosoma armatum (Giard and Barrois, 1874) Cobb, 1929 n. syn.
    Notochaetosoma macrocephalum (Schepotieff, 1907) Cobb, 1929 n. syn.
    Notochaetosoma hibernicum (Southern, 1914) Cobb, 1929 n. syn.
    Drepanonema claparedii (Mechnikov, 1867) Cobb, 1933
    Drepanonema armatum (Giard and Barrois, 1874) Cobb, 1933 n. syn.
    Drepanonema macrocephalum (Schepotieff, 1907) Cobb, 1933 n. syn.
    Drepanonema hibernicum (Southern, 1914) Cobb, 1933 n. syn.
    Claparediella claparedii (Mechnikov, 1867) Filipjev, 1934
    Claparediella macrocephala (Schepotieff, 1907) Filipjev, 1934 n. syn.
    Draconema macrocephalum (Schepotieff, 1908) Schuurmans Stekhoven, 1935 n. syn.
    Draconema hibernicum (Southern, 1914) Kreis, 1938 n. syn.

Measurements (23 ♀♀): L = 1.1 (0.9-1.5) mm; b = 9.1 (7.3-10.2); c = 10.2 (7.7-12.8); V = 54 (51-56)%; CAT = 29 (25-33) μm; SER (L/W) = 2.2 (2.0-2.4); SER (E/L) = 13 (12-15)%; No Large SER Ann = 11 (10-13); SS = 17-86 μm; first SIAT = 44 (39-48) μm; last SIAT = 33 (27-37) μm; first SvAT = 36 (34-39) μm; last SvAT = 22 (20-25) μm; No SIAT = 10 (9-11); No SvAT = 16 (13-18); Non-ann Term to Tail Length = 41 (35-46)%; T/ABD = 5.4 (4.5-6.0).

(23 ♂♂): L = 1.1 (0.9-1.2) mm; a = 23.6 (20.0-27.0); b = 9.4 (7.7-11.2); c = 9.1 (7.7-11.0); CAT = 29 (24-34) μm; SER (L/W) = 2.1 (1.9-2.4); SER (E/L) = 12 (10-13)%; No Large SER Ann = 10 (7-12); SS = 19-92 μm; first SIAT = 47 (37-60) μm; last SIAT = 35 (33-46) μm; first SvAT = 38 (34-44) μm; last SvAT = 23 (21-38) μm; No SIAT = 9 (8-9); No SvAT = 15 (13-17); Non-ann Term to Tail Length = 32 (29-35)%; T/ABD = 4.9 (4.4-5.4); Spic = 74 (66-82) μm; Gub = 19 (17-22) μm.

Males Emended.—Amphids elongate loop-shaped. CAT typical of genus. Longest SS on swollen esophageal region. Long and short setae intermingled with SIAT. Four to 5 long setae, usually 4, with SIAT, setae longer than adjacent tube, not alternating with tubes. Caudal glands extend anterior to anus 2.6 to 3.3 times ABD. Anal setae uniformly tapered; 2 shorter pairs 9 to 11 μm long, and 2 longer pairs 13 to 18 μm long (fig. 14); 2 pairs anterior, 1 almost adjacent to, and 1 posterior to anus. Anal flap short, not crenated. Five pairs of short setae on non-annulated tail region, position measured from last complete tail annule to tail tip; 2 subventral pairs, 1 about 33%, 1 about 50%; 3 subdorsal pairs, 1 just posterior to last complete tail annule, 1 about 25%, and 1 about 66%.

Females Emended.—Similar to males. Amphids usually unispiral, sometimes ventral arm only partially converging on dorsal arm. Vulva encircled with minute spine-like projections, smaller but similar to *D. ophicephalum*. Paravulval setae 8 to 9 μm long. Longest SS on swollen esophageal region. Short setae alternate with SIAT. Caudal glands extend anterior to anus 2.3 to 3.0 times ABD. Anal flap long, crenate. Four pairs of short setae on non-annulated tail region; 3 subdorsal pairs as in males; 1 subventral pair about 50%.

First-Stage Larvae.—L = 0.3 mm. (See Generic Description.)

Second-Stage Larvae.—L = 0.4 to 0.5 mm. Similar to adults. Amphids unispiral, ventral arm beginning to form double spiral at base of dorsal arm. One large annule just posterior to rostrum. CAT and SIAT typical of genus. One short subdorsal pair of setae about 50% on non-annulated tail region.

Third-Stage Larvae.—L = 0.5 to 0.7 mm. Similar to adults. Amphids unispiral, or beginning to form double spiral as in second-stage larvae. One to 3 large annules just posterior to rostrum. CAT and SlAT typical of genus. Two pairs of short setae on non-annulated tail region; 1 subdorsal pair about 50%; 1 lateral pair just posterior to last complete tail annule.

Fourth-Stage Larvae.—L = 0.7 to 0.8 mm. Similar to adults. Amphids unispiral. Three to 4 large annules just posterior to rostrum. CAT and SlAT typical of genus; 7 to 9 VAT, usually 9. Four pairs of setae on non-annulated tail region; 1 subventral pair about 33%; 3 subdorsal pairs, 1 long pair on or just anterior to last complete tail annule equal in length to long setae on annulated tail, 1 short pair just posterior to long subdorsal pair, 1 short pair about 66%.

Type Habitat.—Marine.

Type Locality.—Salerno, Italy. Collected by I. Mechnikov in September, 1866.

Distribution.—Known distribution. Marseille, France; Naples and Salerno, Italy.

This species has also been recorded from the English Channel; Roscoff (Coast of Brittany), France; Clare Island in Clew Bay, Ireland; and Norway.

Diagnosis.—Most closely resembles *D. cephalatum* differs by males with all uniformly tapered anal setae; and females with fewer SlAT. Differs from *D. cephalatum, D. haswelli* and *D. antarcticum* n. sp. by absence of long setae on non-annulated tail region. Differs from *D. chilense* n. sp. and *D. ophicephalum* by fewer SlAT.

Second-stage larvae differ from other known second-stage *Draconema* by ventral arm of amphid beginning to form double spiral. Third-stage most closely resemble third-stage *D. antarcticum* n. sp. differ in smaller size, and 1 short lateral pair of setae just posterior to last tail annule on non-annulated tail region. Differs from third stages of *D. cephalatum, D. haswelli,* and *D. ophicephalum* by absence of long setae on non-annulated tail region. Fourth-stage larvae most closely resemble fourth-stage *D. ophicephalum* differ by absence of 1 lateral pair of setae about 50% on non-annulated tail region. Differs from fourth stages of *D. haswelli* by unispiral amphid; from *D. antarcticum* n. sp. by smaller size; and from *D. cephalatum* and *D. chilense* n. sp. by 1 subventral pair of setae about 33% on non-annulated tail region.

Type specimens of this species were not available. We were able to study specimens collected by S. A. Gerlach from Naples, Italy, which is very close to the type locality. By comparing the original description and illustrations with these specimens, we were able to distinguish *D. claparedii* from the other known species of *Draconema*.

Discussion.—In 1874, A. Giard and J. H. Barrois described *Chaetosoma armatum* from a male collected at Roscoff on the coast of Brittany, France. This species was compared to *C. ophicephalum* Claparéde, 1863 and *C. claparedii*, and concluded it was closer to the latter species. From the original description and illustrations, we conclude it is the same as *D. claparedii*.

A. Schepotieff in 1907 and 1908 described *Chaetosoma macrocephalum* from a third-stage larva and a male collected at Naples and Bergen, Italy. The male of *C. macrocephalum* described in 1908 appears to be the same as *D. claparedii*, therefore, we propose that *C. macrocephalum* is a synonym of *D. claparedii*.

In 1914, R. Southern described *Chaetosoma hibernicum* from Clare Island in Clew Bay, Ireland. From the original description and illustrations, and on the basis of specimens examined of *D. claparedii* collected from the Mediterranean, *C. hibernicum* is also judged a synonym of *D. claparedii*.

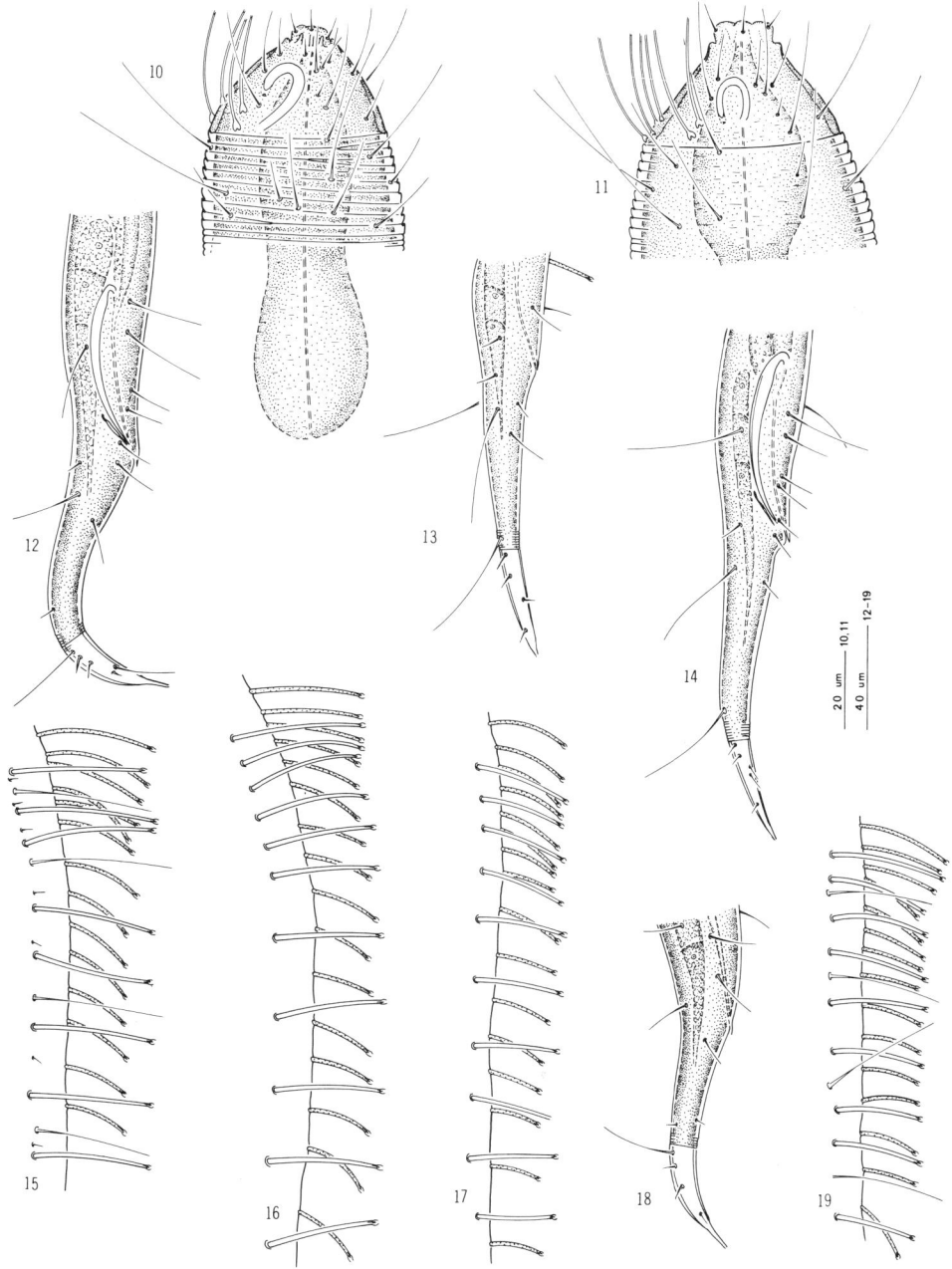

2. Figs. 10, 12, 15–16, 18: *Draconema haswelli* Irwin-Smith. 10) Male, head and esophageal region, correct number of CAT not illustrated; 12) Male, tail; 15) Male, PAT; 16) Female, PAT[a]; 18) Female, tail. Figs. 11, 13-14, 17, 19: *Draconema claparedii* Mechnikov. 11) Male, head; 13) Female, tail; 14) Male, tail; 17) Female, PAT[a]; 19) Male, PAT[a].

a. Short SS with SlAT not illustrated.

Draconema ophicephalum (Claparède, 1863) Filipjev, 1918

(Figs. 20 to 27)

Syn: *Chaetosoma ophicephalum* Claparède, 1863
Chaetosoma annulatum Ditlevsen, 1915 n. syn.
Draconema cephalatum Steiner, 1916 *nec.* Cobb, 1913 n. syn.
Draconema cephalatum Filipjev, 1918 *nec.* Cobb, 1913 n. syn.
Draconema ponticum Filipjev, 1918 n. syn.
Draconema micoletzkyi Kreis, 1928 n. syn.
Notochaetosoma ophicephalum (Claparède, 1863) Cobb, 1929
Notochaetosoma annulatum (Ditlevsen, 1915) Cobb, 1929 n. syn.
Drepanonema ophicephalum (Claparède, 1863) Cobb, 1933
Drepanonema annulatum (Ditlevsen, 1915) Cobb, 1933 n. syn.
Claparediella ophicephala (Claparède, 1863) Filipjev, 1934
Claparediella annulatum (Ditlevsen, 1915) Filipjev, 1934 n. syn.
Draconema annulatum (Ditlevsen, 1915) Schuurmans Stekhoven, 1935 n. syn.
Tristicochaeta ponticum (Filipjev, 1918) Cobb, 1935 n. syn.
Tristicochaeta micoletzkyi (Kreis, 1928) Cobb, 1935 n. syn.

Measurements (5 ♀♀): L = 1.5 (1.2–1.7) mm; b = 11.1 (9.8–11.7); c = 11.1 (7.4–13.5); V = 56 (54–57)%; CAT = 33 (29–36) µm; SER (L/W) = 1.8 (1.7–1.9); SER (E/L) = 10 (10–11)%; No Large SER Ann = 9 (9–10); SS = 21–62 µm; first SlAT = 45 (41–49) µm; last SlAT = 38 (33–41) µm; first SvAT = 43 (38–46) µm; last SvAT = 23 (21–26) µm; No SlAT = 20 (18–22); No SvAT = 19 (18–20); Non-ann Term to Tail Length = 43 (38–46)%; T/ABD = 4.7 (4.2–5.2).

(3 ♂♂): L = 1.3 (1.2–1.6) mm; a = 22.4 (16.7–28.6); b = 10.5 (10.0–11.4); c = 9.6 (8.7–10.0); CAT = 37 (37–38) µm; SER (L/W) = 1.7 (1.6–1.8); SER (E/L) = 10 (10–11)%; No Large SER Ann = 11 (9–12); SS = 22–60 µm; first SlAT = 50 (49–51) µm; last SlAT = 41 (40–44) µm; first SvAT = 41 (37–44) µm; last SvAT = 24 (23–26) µm; No SlAT = 15 (14–16); No SvAT = 18 (18–19); Non-ann Term to Tail Length = 30 (30–31)%; T/ABD = 4.5 (4.2–4.7); Spic = 89 (84–92) µm; Gub = 22 (21–23) µm.

Males Emended.—Amphids elongate loop-shaped. CAT typical of genus. Longest SS on swollen esophageal region. Long and short setae intermingled with SlAT. Four long setae with SlAT, setae longer than adjacent tube, not alternating with tubes. Caudal glands extend anterior to anus 2.9 to 3.7 times ABD. Anal setae uniformly tapered (fig. 24), 14 to 20 µm long, 2 pairs anterior and 2 posterior to anus. Anal flap short, slightly crenated. Six pairs of setae on non-annulated tail region, position measured from last complete tail annule to tail tip; 1 lateral pair about 66%, on some specimens slightly dorsal; 3 subdorsal pairs, 1 long pair just posterior to last complete tail annule or 1 to 2 annule widths just anterior to last tail annule equal in length to long setae on annulated tail region, 1 short pair just posterior to long subdorsal pair, 1 short pair about 33%; 2 subventral pairs, 1 about 33%, 1 about 50%.

Females Emended.—Similar to males. Vulva encircled with prominent, spine-like projections (figs. 22 to 23). Paravulval setae 15 to 20 µm long. Longest SS on mid-body region. Short setae intermingled with SlAT. Caudal glands extend anterior to anus 2.6 to 5.1 times ABD. Anal flap long, slightly crenated. Four to 5 pairs of setae on non-annulated tail region; 1 subventral pair about 66%; 3 to 4 subdorsal pairs, 1 long pair just posterior to last complete tail annule equal in length to long setae on annulated tail region, 1 short pair about 25%, 1 short pair about 33%, 1 short pair present or absent about 75%.

First-Stage Larvae.—Not observed.

Second-Stage Larvae.—L = 0.4 to 0.5 mm. Similar to adults. Amphids unispiral. One to 2 large annules posterior to rostrum. CAT and SlAT typical of genus. One subdorsal pair of short setae about 50% on non-annulated tail region.

Third-Stage Larvae.—L = 0.5 to 0.7 mm. Similar to adults. Amphids unispiral. One to 3 large annules posterior to rostrum. CAT and SlAT typical of genus. Three pairs of setae on non-annulated tail region; 1 long subdorsal pair just posterior to last complete tail annule or 2 to 4 annule widths anterior to last tail annule equal in length to long setae on annulated tail region; 1 short subdorsal pair about 50%; 1 lateral pair just posterior to last tail annule.

Fourth-Stage Larvae.—L = 0.7 to 1.0 mm. Similar to adults. Amphids unispiral. Three to 4 large annules posterior to rostrum. CAT and SlAT typical of genus; 9 VAT. Four to five pairs of setae on non-annulated tail region; 1 lateral pair about 50%; 1 subventral pair about 33%; 2 to 3 subdorsal pairs, 1 long pair just posterior to last complete tail annule equal in length to long setae on annulated tail region, 1 short pair about 25%, 1 short pair present or absent about 75%.

Type Habitat.—Marine.

Type Locality.—St. Vaast (Coast of Normandy), France.

Distribution.—Known distribution. Marseille and St. Vaast (Coast of Normandy), France; Cala Cupa and Naples, Italy.

This species has also been reported from the Barents Sea; Danish Belt Sea; English Channel; Greenland; Bergen, Italy; Kiel Bay, North Sea; Norway; Georgievskii Monastery, Kruglaya Bay, and Sevastopol, Union of Soviet Socialist Republics.

Diagnosis.—Most closely resembles *D. chilense* n. sp. differs by larger size; and longer tails, last SlAT, first SvAT; males by uniformly tapered anal setae, and 1 long pair of setae on non-annulated tail region; females by longer paravulval setae, and shorter non-annulated tail region. Differs from *D. cephalatum*, *D. haswelli*, *D. claparedii* and *D. antarcticum* n. sp. by greater number of SlAT.

Second-stage larvae most closely resemble second-stage *D. antarcticum* n. sp. differ by smaller size; smaller unispiral amphids, never elongate loop-shaped; and fewer large annules posterior to rostrum. Differs from second stages of *D. cephalatum* and *D. haswelli* by 1 pair of setae on non-annulated tail region; from *D. claparedii* by unispiral amphid. Third-stage larvae most closely resemble third-stage *D. haswelli* differ by 1 lateral pair of setae just posterior to last complete tail annule on non-annulated tail region. Differ from *D. claparedii* and *D. antarcticum* n. sp. by 1 long pair of setae on non-annulated tail region; and from *D. cephalatum* by absence of setae near tail tip. Fourth-stage larvae most closely resemble fourth-stage *D. claparedii* differ by 1 lateral pair of setae about 50% on non-annulated tail region. Differ from *D. haswelli* by unispiral amphid, and from *D. antarcticum* n. sp. by smaller size and 1 long pair of setae on non-annulated tail region; from *D. cephalatum* and *D. chilense* n. sp. by 1 subventral pair of setae about 33% on non-annulated tail region.

Type specimens of this species were not available. We examined specimens collected by S. A. Gerlach from Naples, Italy, and specimens collected from Marseille, France. These specimens combined with the original description and illustrations enabled us to distinguish this species.

Discussion.—In 1915, H. Ditlevsen described *Chaetosoma annulatum* from Greenland. The number of posterior adhesion tubes in the male is not given in the original description but based upon the number of adhesion tubes in the female (22), we conclude this species to be a synonym of *D. ophicephalum*.

G. Steiner in 1916, described specimens he considered to be *D. cephalatum* Cobb, 1913, from the Barents Sea. From his original description and illustrations of these specimens we consider them to be *D. ophicephalum*.

In 1918, I. N. Filipjev redescribed a nema he considered to be *D. cephalatum* Cobb, 1913, from the Black Sea near Georgievskii Monastery and Kruglaya Bay. On the basis of the number of posterior adhesion tubes in the male and female, we conclude this nematode to be *D. ophicephalum*.

In this same publication (1918) Filipjev described *D. ponticum* from a fourth-stage larva collected from near the Georgievskii Monastery. On the basis of our knowledge of the morphology and stable taxonomic characters of immature forms we are synonymizing *D. ponticum* with *D. ophicephalum*.

H. A. Kreis in 1928 described *D. micoletzkyi* from Naples, Italy. In his original description he compares *D. micoletzkyi* with *D. cephalatum* Cobb, 1913 and concluded they were different species. Based on his original description and illustrations we judge *D. micoletzkyi* to be a synonym of *D. ophicephalum*.

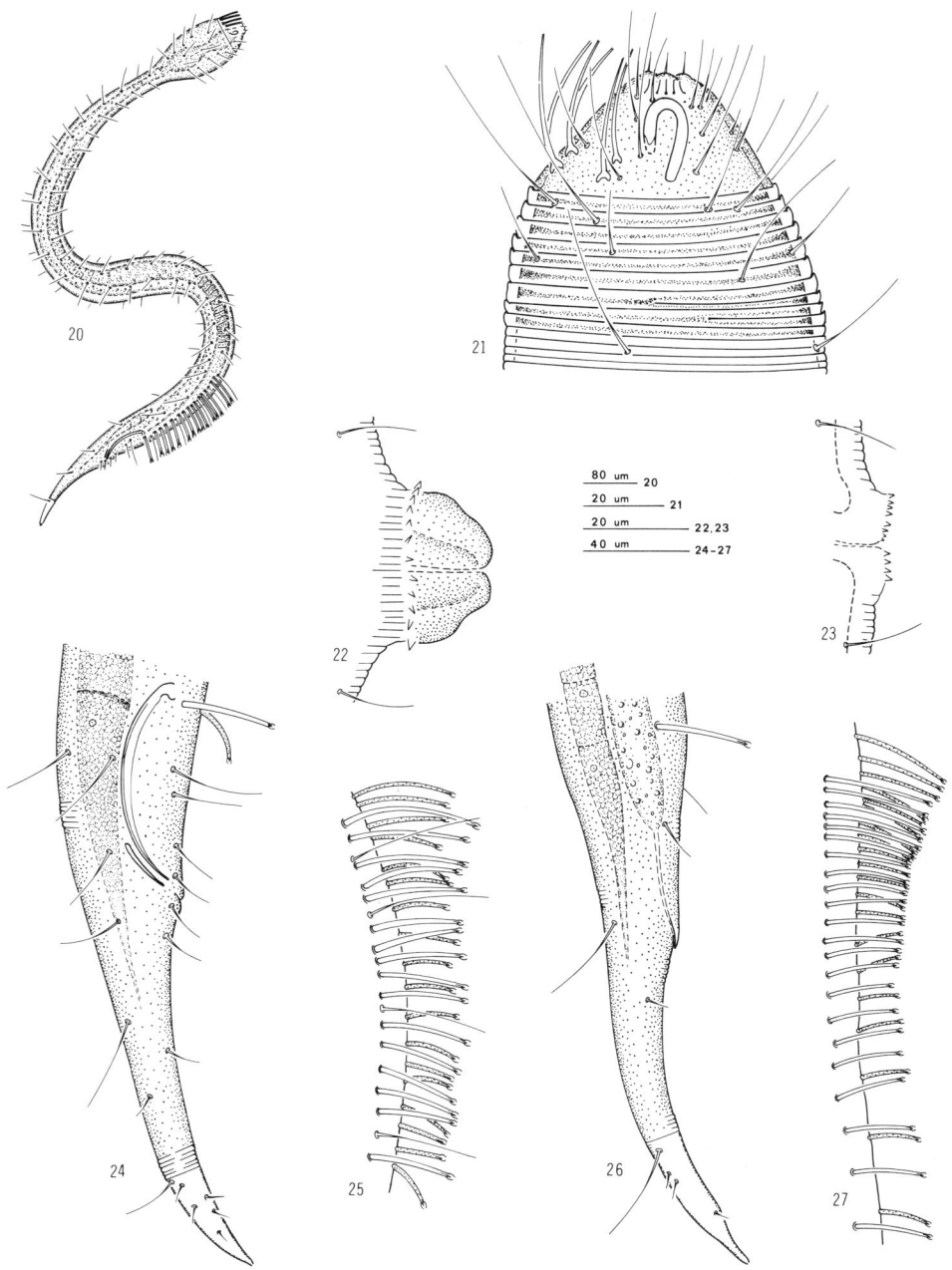

3. Figs. 20–27: *Draconema ophicephalum* Claparéde. 20) Male, full length[a], SvAT and correct number of CAT not illustrated; 21) Male, head, shows CAT location on rostrum; 22–23) Female, extruded and normal vulva, paravulval setae and spine-like projections encircling the vulva; 24) Male, tail; 25) Male, PAT[b]; 26) Female, tail; 27) Female, PAT[b].

a. Total number and position of setae on non-annulated tail not illustrated.
b. Short SS with SlAT not illustrated.

Draconema antarcticum n. sp.
(Figs. 9, 28 to 41)

Measurements (23 ♀♀): L = 1.7 (1.3–2.1) mm; b = 10.9 (8.8–12.6); c = 11.3 (9.3–14.3); V = 56 (53–59)%; CAT = 36 (30–41) μm; SER (L/W) = 1.8 (1.6–2.2); SER (E/L) = 10 (9–12)%; No Large SER Ann = 11 (9–12); SS = 23–78 μm; first SlAT = 69 (55–76) μm; last SlAT = 57 (54–62) μm; first SvAT = 62 (54–73) μm; last SvAT = 38 (33–43) μm; No SlAT = 15 (14–16); No SvAT = 17 (16–19); Non-ann Term to Tail Length = 44 (36–51)%; T/ABD = 5.4 (4.2–6.3).

(24 ♂♂): L = 1.7 (1.2–2.2) mm; a = 22.4 (16.1–29.3); b = 11.3 (8.9–13.8); c = 10.3 (7.9–13.8); CAT = 38 (35–43) μm; SER (L/W) = 1.8 (1.6–2.2); SER (E/L) = 10 (9–11)%; No Large SER Ann = 11 (9–14); SS = 17–77 μm; first SlAT = 78 (70–86) μm; last SlAT = 61 (48–67) μm; first SvAT = 71 (62–80) μm; last SvAT = 41 (37–45) μm; No SlAT = 9 (9–10); No SvAT = 16 (15–18); Non-ann Term to Tail Length = 33 (29–41)%; T/ABD = 4.7 (3.9–5.4); Spic = 94 (83–104) μm; Gub = 31 (24–35) μm.

(Holotype ♂): L = 1.7 mm; a = 22.6; b = 12.3; c = 10.8; CAT = 37 μm; SER (L/W) = 1.8; SER (E/L) = 10%; No Large SER Ann = 10; SS = 25–59 μm; first SlAT = 71 μm; last SlAT = 60 μm; first SvAT = 62 μm; last SvAT = 40 μm; No SlAT = 9; No SvAT = 15; Non-ann Term to Tail Length = 32%; T/ABD = 4.8; Spic = 90 μm; Gub = 24 μm.

Male Holotype.—(Figures in parentheses refer to range within species.) Amphids elongate loop-shaped. CAT typical of genus. Longest SS on swollen esophageal region. Long and short setae intermingled with SlAT. Four (3 to 4) long setae with SlAT, setae longer than adjacent tube, not alternating with tubes. Caudal glands extend anterior to anus 2.6 (2.3 to 3.8) times ABD. Anal setae uniformly tapered, 26 to 31 μm (23 to 31 μm) long, 2 pairs anterior and 2 posterior to anus, positioned more laterally than other known *Draconema* species (fig. 32). Anal flap short, not crenated. Five pairs of setae on non-annulated tail region, position measured from last complete tail annule to tail tip; 1 long subventral pair about 50% equal in length to long setae on annulated tail region; 1 lateral pair adjacent to long subventral pair; 3 subdorsal pairs, 1 just posterior to last complete tail annule (absent on some specimens), 1 about 13%, 1 about 66%.

Females.—Similar to males. Vulva encircled with minute, spine-like projections, not as conspicuous as other *Draconema*. Paravulval setae 11 to 14 μm long. Longest SS on swollen esophageal region. Short setae intermingled with SlAT. Caudal glands extend anterior to anus 2.8 to 3.6 times ABD. Four to 5 pairs of setae on non-annulated tail region; 1 lateral pair about 75%; 1 subventral pair present or absent about 66%; 3 subdorsal pairs, 1 long pair just posterior to last complete tail annule or 1 to 3 annule widths anterior to last annule equal in length to long setae on annulated tail region, 1 short pair about 25%, 1 short pair about 75%.

First-Stage Larvae (Figs. 33, 36, 38).—L = 0.3 to 0.4 mm. (See Generic Description.)

Second-Stage Larvae (Figs. 34, 39).—L = 0.5 to 0.6 mm. Similar to adults. Amphids elongate loop-shaped to circular unispiral. Two to 5 large annules posterior to rostrum. CAT and SlAT typical of genus. One pair of short subdorsal setae present or absent, usually present, about 50% on non-annulated tail region.

Third-Stage Larvae (Figs. 35, 40).—L = 0.8 to 0.9 mm. Similar to adults. Amphids partially closed elongate loop-shaped, or unispiral. Three to 5 large annules posterior to rostrum. CAT and SlAT typical of genus. Two pairs of short setae on non-annulated tail region; 1 subdorsal pair about 50%; 1 lateral or slightly subventral pair, if lateral about 13%, if subventral just posterior to last complete tail annule or 13%.

Fourth-Stage Larvae (Figs. 37, 41).—L = 1.2 to 1.4 mm. Similar to adults. Amphids elongate loop-shaped, unispiral or with ventral arm beginning to form double spiral against dorsal arm. Four to 7 large annules just posterior to rostrum. CAT and SlAT typical of genus; 8 to 10 VAT, usually 9. Three to 4 pairs of short setae on non-annulated tail region; 1 lateral pair about 50%; 1 subventral pair present or absent about 33%; 2 subdorsal pairs, 1 just posterior to last complete tail annule or 1 to 2 annule widths just anterior to last annule, 1 about 66%.

Holotype (♂).—Collected January 16, 1970 by R. W. Timm and D. R. Viglierchio. U. S. National Museum of Natural History, Smithsonian Institution, Washington, D. C., No. 52004.

Paratypes.—27 ♀♀, 19 ♂♂. Same data as holotype. Thirteen females and 9 males deposited at U. S. National Museum of Natural History, Smithsonian Institution, Washington, D. C.; and 14 females and 10 males deposited University of California, Davis (UCNC 1456 to 1461).

Larval Stages.—2 first, 3 second, 10 third, 21 fourth. Same data as holotype. Deposited in same nematode collections as paratypes.

Type Habitat.—Marine, at 540 meters.

Type Locality.—Scott Base, Antarctica.

Distribution.—Cape Royds, McMurdo Sound, and Scott Base, Antarctica.

Diagnosis.—Most closely resembles *D. haswelli*; males differ by longer CAT, first SIAT and SvAT, last SvAT, spicules, gubernaculum, and SS opposite PAT (*D. antarcticum* = 31 to 49 μm; *D. haswelli* = 15 to 27 μm); and absence of 1 long subdorsal pair of setae just posterior to last complete tail annule on non-annulated tail region. Females differ by greater number of SIAT, longer last SIAT, shorter caudal glands, and amphids always elongate loop-shaped. Males differ from *D. cephalatum* and *D. claparedii* by 1 long subventral pair of setae on non-annulated tail; and from *D. chilense* n. sp. and *D. ophicephalum* by fewer SIAT. Females differ from *D. claparedii* by 1 long pair of setae on non-annulated tail region and greater number of SIAT; and from *D. cephalatum* by amphids elongate loop-shaped never unispiral, and greater number of SIAT. Females differ from *D. chilense* n. sp. and *D. ophicephalum* by fewer SIAT.

Second-stage larvae most closely resemble second-stage *D. ophicephalum* differ by greater number of large annules posterior to rostrum, and larger size. Differ from second-stage larvae of *D. cephalatum* by 1 pair of setae on non-annulated tail region; from *D. claparedii* by unispiral amphids, and from *D. haswelli* by larger size. Third-stage larvae most closely resemble third-stage *D. claparedii* differ by larger size. Differ from other known third-stage *Draconema* by absence of long setae on non-annulated tail region. Fourth-stage larvae differ from other known fourth-stage *Draconema* by larger size.

Draconema chilense n. sp.

(Figs. 42 to 48)

Measurements (15 ♀♀): L = 0.8 (0.6-1.1) mm; b = 7.6 (6.0-9.8); c = 10.3 (7.8-13.5); V = 55 (50-59)%; CAT = 24 (19-30) μm; SER (L/W) = 1.7 (1.5-1.9); SER (E/L) = 14 (11-17)%; No Large SER Ann = 10 (8-11); SS = 13-46 μm; first SIAT = 38 (33-41) μm; last SIAT = 30 (28-32) μm; first SvAT = 33 (29-37) μm; last SvAT = 20 (16-21) μm; No SIAT = 22 (20-24); No SvAT = 18 (15-20); Non-ann Term to Tail Length = 56 (51-60)%; T/ABD = 3.6 (3.2-3.9).

(20 ♂♂): L = 0.9 (0.7-1.1) mm; a = 13.9 (12.2-16.5); b = 8.4 (7.1-10.2); c = 8.7 (6.9-10.7); CAT = 25 (14-28) μm; SER (L/W) = 1.8 (1.5-1.9); SER (E/L) = 13 (10-15)%; No Large SER Ann = 10 (9-12); SS = 12-40 μm; first SIAT = 39 (35-42) μm; last SIAT = 30 (26-33) μm; first SvAT = 31 (27-36) μm; last SvAT = 20 (15-23) μm; No SIAT = 17 (15-19); No SvAT = 16 (14-18); Non-ann Term to Tail Length = 36 (30-38)%; T/ABD = 3.4 (3.1-3.8); Spic = 101 (93-108) μm; Gub = 30 (21-33) μm.

(Holotype ♂): L = 1.0 mm; a = 13.0; b = 9.6; c = 6.9; CAT = 27 μm; SER (L/W) = 1.7; SER (E/L) = 10%; No Large SER Ann = 10; SS = 19-36 μm; first SIAT = 38 μm; last SIAT = 30 μm; first SvAT = 30 μm; last SvAT = 18 μm; No SIAT = 16; No SvAT = 15; Non-ann Term to Tail Length = 36%; T/ABD = 3.4; Spic = 108 μm; Gub = 28 μm.

Male Holotype.—(Figures in parentheses refer to range within species.) Amphids elongate loop-shaped. CAT typical of genus. Longest SS on mid-body region (on other specimens longest usually on esophageal region). Long and short setae intermingled with SIAT. Four long setae with SIAT, setae longer than adjacent tube, not alternating with tubes. Caudal glands extend anterior to anus 3.0 (2.7 to 3.7) times ABD. Anal setae short, prominent, broad-based, unevenly tapered, 8 to 9 μm (6 to 11 μm) long; 1 pair anterior, 1 adjacent to, and 2 posterior to anus (fig. 48). Anal flap short, not crenated. Five (4 to 5) pairs of short setae on non-annulated tail region, position measured from last complete tail annule to tail tip; 1 lateral pair (absent on some specimens) about 66%; 2 subdorsal pairs, 1 just posterior to last complete tail annule, and 1 about 33%; 2 subventral pairs adjacent to each other (on other specimens sometimes inner pair slightly posterior) about 50%.

Females.—Similar to males. Without spine-like projections encircling vulva. Paravulval setae 10 to 13 μm long. Longest SS on mid-body region. Short setae intermingled with SIAT. Caudal glands extend anterior to anus 2.6 to 4.1 times ABD. Anal flap long, not crenated. Four to 6 pairs of setae on non-annulated tail region; 1 lateral pair about 66%; 3 subdorsal pairs, 1 long pair just

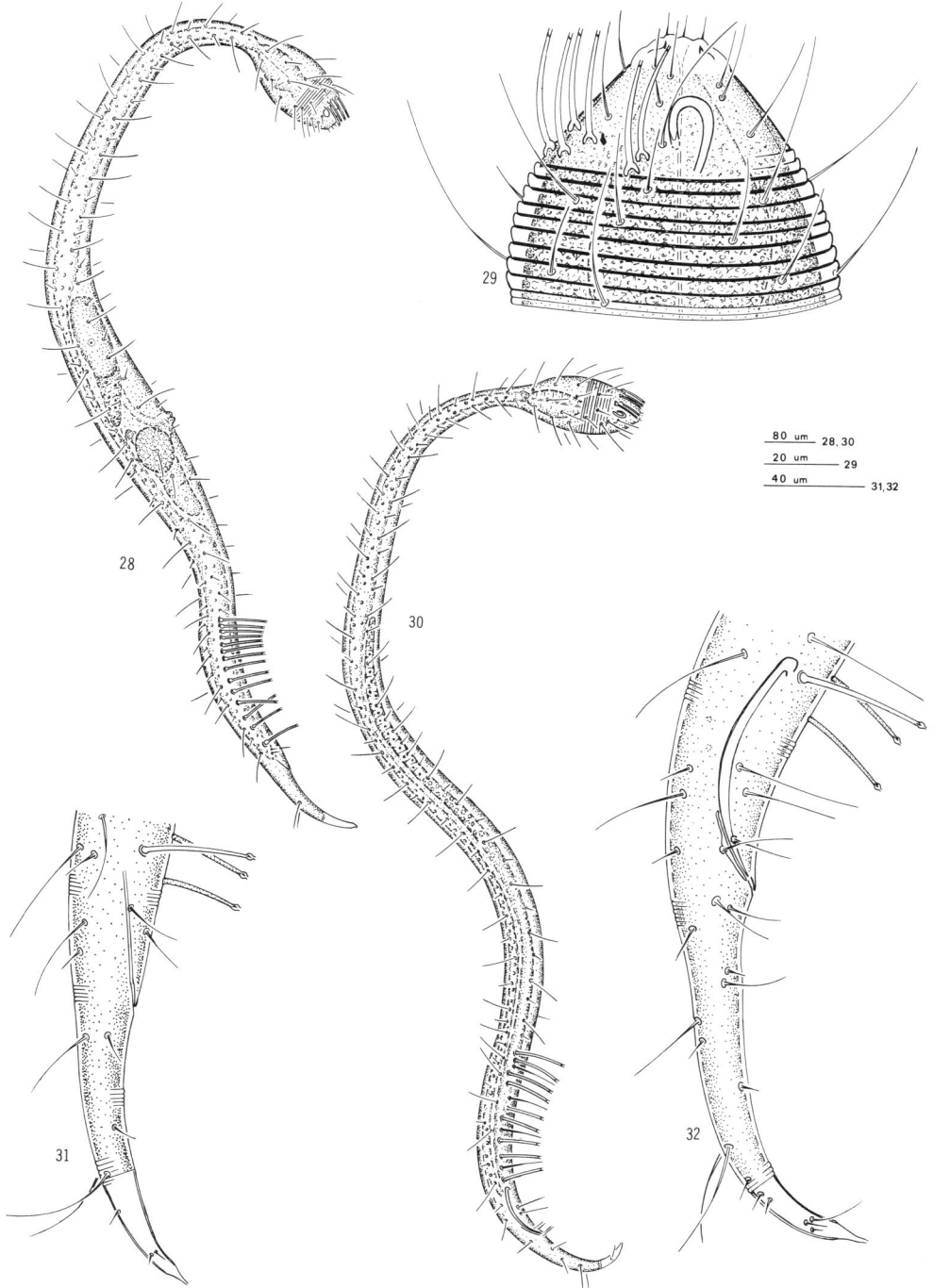

4. Figs. 28–32: *Draconema antarcticum* n. sp. 28) Female, full length[a], SvAT and correct number of CAT not illustrated; 29) Male, head; 30) Male, full length[a], SvAT and correct number of CAT not illustrated; 31) Female, tail; 32) Male, tail.

[a]. Total number and position of setae on non-annulated tail region and short SS with SlAT not illustrated.

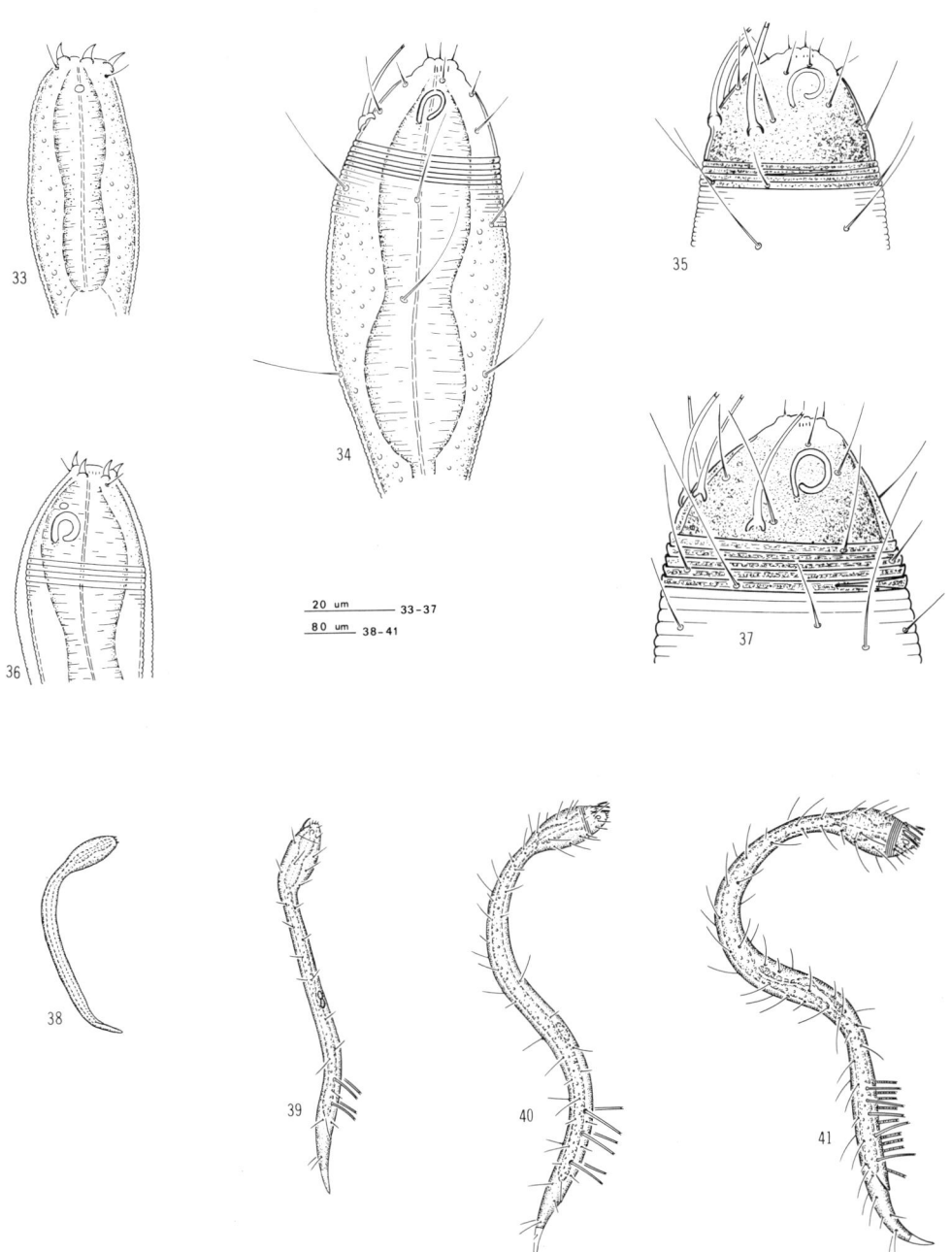

5. Figs. 33–41: *Draconema antarcticum* n. sp. 33) First-stage larva, head and esophageal region; 34) Second-stage larva, head and esophageal region; 35) Third-stage larva, head; 36) First-stage larva, molting, head and esophageal region; 37) Fourth-stage larva, head; 38) First-stage larva, full length; 39) Second-stage larva, full length[a]; 40) Third-stage larva, full length[a]; 41) Fourth-stage larva, full length[a].

a. Total number and position of setae on non-annulated tail region not illustrated.

posterior to last complete tail annule equal in length to long setae on annulated tail region, 1 short pair just posterior to long subdorsal pair, 1 short pair about 50%; 2 subventral pairs present or absent, usually 1 just anterior to lateral pair, sometimes 1 just posterior to last complete tail annule.

First-, Second-, and Third-Stage Larvae.—Not observed.

Fourth-Stage Larvae.—L = 0.5 to 0.6 mm. Similar to adults. Amphids unispiral or elongate loop-shaped. Two to 5 large annules just posterior to rostrum. CAT and SlAT typical of genus; 9 VAT. Four pairs of setae on non-annulated tail region; 1 lateral pair about 50%; 1 long subdorsal pair just posterior to last complete tail annule equal in length to long setae on annulated tail region; 2 subdorsal to sublateral pairs, 1 just posterior to long subdorsal pair, 1 just posterior to latter pair.

Holotype (♂).—Collected July 17, 1966, by Robert Woollacott. Catalogue No. UCNC 1416, University of California, Davis.

Paratypes.—7 ♀♀, 12 ♂♂. Same data as holotype. Paratypes deposited at University of California, Davis (UCNC 1462 to 1463); USDA Nematode Collection, Beltsville, Maryland; U. S. National Museum of Natural History, Smithsonian Institution, Washington, D. C.; and Laboratoria voor Morfologie en Systematiek, Museum voor Dierkunde, Gent, Belgium.

Larval Stages.—1 fourth. Same data as holotype. Deposited at University of California, Davis.

Type Habitat.—Marine, associated with Coralline Red Algae.

Type Locality.—Mas A Tierra Island, Juan Fernandez Islands, Chile.

Distribution.—Mas A Tierra Island, Juan Fernandez Islands and Puerto Vagageria, Chile.

Diagnosis.—Males differ from other known *Draconema* by 4 pairs of short, prominent, broad-based, unevenly tapered anal setae. Females most closely resemble *D. ophicephalum* differ by smaller size, shorter tail, absence of spine-like projections encircling vulva, shorter paravulval setae, shorter last SlAT and first SvAT, and longer non-annulated tail region. Females differ from *D. cephalatum, D. haswelli, D. claparedii,* and *D. antarcticum* n. sp. by greater number of SlAT.

Fourth-stage larvae differ from other known fourth-stage *Draconema* by smaller size. Most closely resemble fourth-stage *D. haswelli* differ by 2 subdorsal to sublateral pairs of setae on non-annulated tail region. Differ from fourth-stage larvae of *D. antarcticum* n. sp. by 1 long pair of setae on non-annulated tail region; from *D. ophicephalum* by 4 pairs of setae on non-annulated tail region; and from *D. cephalatum* and *D. claparedii* by 1 lateral pair of setae about 50% on non-annulated tail region.

Genus *Paradraconema*[5] n. gen.

Diagnosis: Draconematinae. Nematodes 0.6 to 1.5 mm long. Body shape as in *Draconema*. Swollen esophageal region generally longer than in *Draconema,* averaging 14% total body length. Rostrum sometimes ornamented with subcuticular markings, setae present. Amphids large, conspicuous, lateral on rostrum; elongate loop-shaped, ventral arm usually longer than dorsal, unispiral, elongate or circular doubled spiral. Twelve CAT, paired, in 2 transverse rows on rostrum. One or 2 pairs of sublateral cephalic acanthiform setae (Ceph Acan-set), slightly ventral on rostrum. When 1 pair, Ceph Acan-set on posterior $1/2$ of rostrum. When 2 pairs of Ceph Acan-set; posterior pair larger, located either about mid-rostrum or on posterior $1/3$ of rostrum; anterior pair smaller, located on anterior $1/2$ of rostrum. Eyespots near or in rostral region, conspicuous or inconspicuous. First 5 to 10 annules sometimes distinct from succeeding annules by subcuticular markings, vacuoles, granulation, and anteriorly directed margins. Remaining body annules with or without subcuticular markings and vacuoles, some species with longitudinally areolated annules. Some species with annules posterior to PAT ornamented and coarser than preceding annules. Some species have minute, anteriorly directed cuticular spines on swollen esophageal region (fig. 94). Paravulval setae present or absent. All PAT anterior to anus, paired or unpaired, tubes variable in number between species. SlAT with non-alternating long and short tubes, differences in length sometimes conspicuous. Males with 2 to 4 pairs, usually 3, uniformly tapered anal setae. Males with or without preanal acanthiform (Acan-set)

---

5. *Paradraconema*, Gr. n., near *Draconema*.

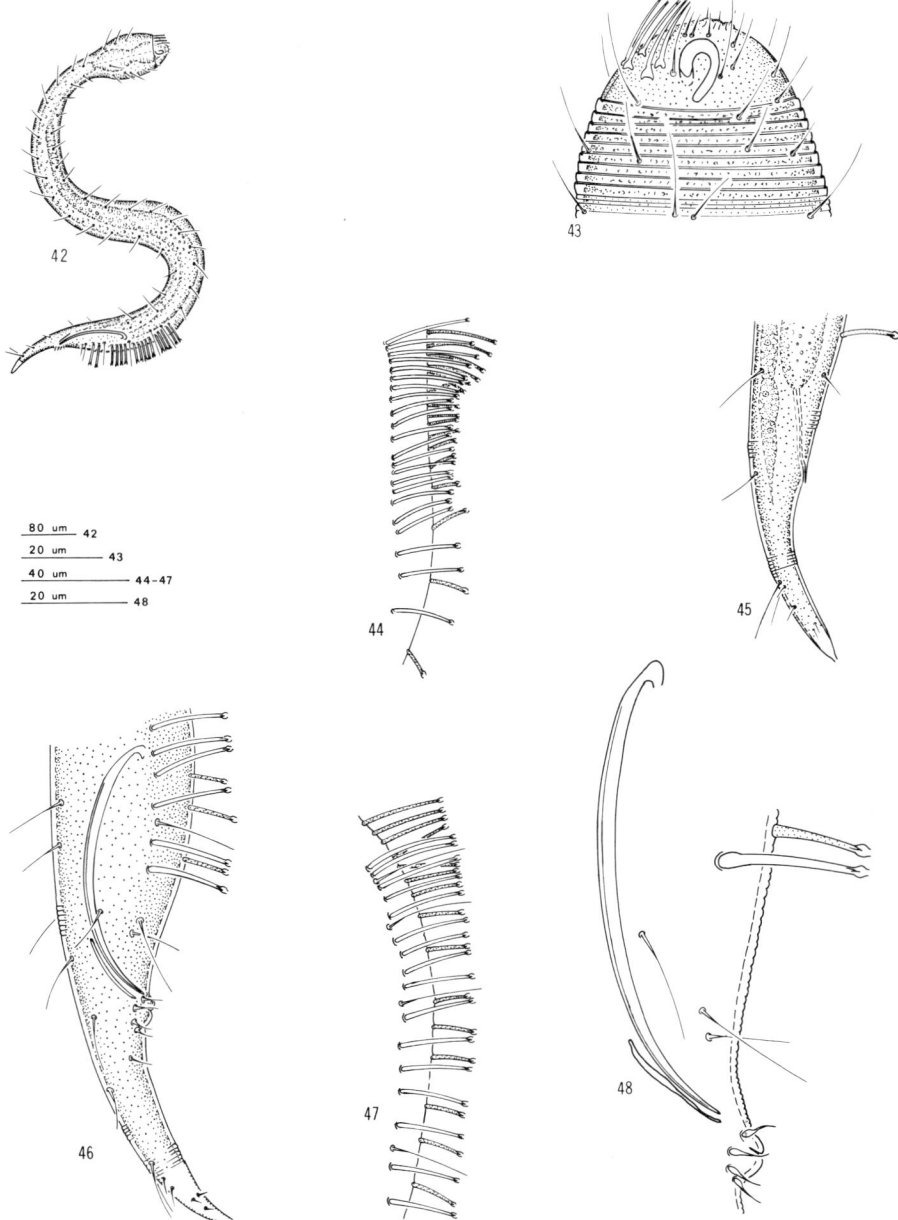

6. Figs. 42–48: *Draconema chilense* n. sp. 42) Male, full length[a], SvAT and correct number of CAT not illustrated; 43) Male, head, correct number of CAT not illustrated; 44) Female, PAT[b]; 45) Female, tail; 46) Male, tail; 47) Male, PAT[b]; 48) Male, anal region.

a. Total number and position of setae on non-annulated tail region not illustrated.
b. Short SS with SlAT not illustrated.

or corniform (Corn-set) setae; either subventral and paired (fig. 80), or ventral and single (fig. 91). Males and females with or without anal flap, sometimes flap crenated (fig. 59). Setae on non-annulated tail region variable in number and location.

Type species: *Paradraconema floridense* n. sp.

Differs from other known Draconematidae by eyespots; and 1 or 2 paired sublateral cephalic acanthiform setae on rostrum (except *Dracograllus stekhoveni* n. sp. and *Paradraconema* differs by 12 CAT). Differs from *Draconema* Cobb, 1913 by absence of prominently enlarged annules on swollen esophageal region. Some males differ from *Draconema* Cobb, 1913 and *Dracograllus* n. gen. by preanal acanthiform or corniform setae. Differs from *Dracograllus* n. gen. by absence of any SS borne on conspicuous, raised, cuticular pedicels (pedicel-setae), and absence of SlAT posterior to anus. Differs from *Dracotoranema* n. gen. by absence of conspicuous alternating long and short, slender, adhesion tubes in rows of SlAT; absence of large ventral corniform seta anterior to SvAT in males.

Larval Stages: No first-stage larvae available in *Paradraconema*. Second-stage larvae 0.4 to 0.8 mm long. Similar to adults. Relaxed body either nearly straight or arched ventrally and dorsally. Esophageal region swollen, rest of body nearly cylindrical. Swollen esophageal region slightly longer than in second-stage *Draconema*. Rostrum sometimes ornamented, setae present. Amphids large, conspicuous, lateral on rostrum, unispiral, or doubled spiral ($^1/_4$ spiral doubled). Single dorsal CAT on rostrum. Ceph Acan-set absent. Eyespots present or absent, located as in adults. Annulation as in adults; ornamentation, if present, obscure. Six rows of SS on swollen esophageal region, 4 sublateral, 1 ventral and 1 dorsal; setae in 4 sublateral rows on remainder of body. Two pairs of SlAT in 2 longitudinal rows. Anal flap sometimes crenated. Non-annulated tail region with obscure punctations, setae variable in number and location. Differentiated from other 2 stages by 1 CAT, 2 pairs of SlAT in 2 rows.

Third-stage larvae 0.4 to 1.0 mm long. Similar to adults. Esophageal region swollen, rest of body nearly cylindrical. Swollen esophageal region slightly longer than in third-stage *Draconema*. Amphids large, conspicuous, lateral on rostrum, ventral arm usually longer than dorsal arm; elongate loop-shaped, unispiral, or double spiral ($^1/_4$ to $^1/_2$ spiral doubled). Three CAT, 2 subdorsal and 1 dorsal, in 2 transverse rows on rostrum. One pair of sublateral Ceph Acan-set slightly ventral on rostrum, located similar to large Acan-set in adults, 1 to 4 μm long. Larvae in species with 2 pairs of Ceph Acan-set sometimes with second minute pair, located as in adults. Eyespots present or absent, if present, obscure, located as in adults. First 2 to 3 annules sometimes distinct with faint to prominent subcuticular ornamentation or anteriorly directed margins; remaining annulation as in adults. Eight rows of SS on swollen esophageal region, 4 sublateral, 2 subdorsal and 2 subventral. Five rows of SS on mid-body region, 4 sublateral and 1 dorsal; 4 sublateral rows on annulated tail region. Some rows with alternating long and short setae. Three pairs of SlAT in 2 longitudinal rows. Non-annulated tail region with punctations, setae variable in number and location. Differentiated from other 2 stages by 3 CAT, 3 pairs of SlAT in 2 rows.

Fourth-stage larvae 0.5 to 1.2 mm long. Similar to adults. Esophageal region swollen, slightly longer than in fourth-stage *Draconema;* young females swollen at mid-body. Amphids large, conspicuous, lateral on rostrum, ventral arm usually longer than dorsal arm; elongate loop-shaped, unispiral, or double spiral ($^1/_2$ to $^3/_4$ spiral doubled). Four CAT, 2 sublateral and 2 subdorsal, in 2 transverse rows on rostrum. Number and location of Ceph Acan-set as in adults, large Acan-set 2 to 5 μm long, if present small Acan-set 1 to 2 μm long. Eyespots located as in adults. First 2 to 7 body annules sometimes distinct with subcuticular ornamentation or with anteriorly directed margins, remaining annulation as in adults. Eight rows of SS on swollen esophageal region, 4 sublateral, 2 subdorsal and 2 subventral. Seven rows of SS on mid-body region, 4 sublateral, 2 subdorsal and 1 ventral; 4 sublateral rows on annulated tail region. Some rows with alternating long and short setae. Three longitudinal rows of PAT, 2 sublateral, 1 ventral. SlAT with 5 pairs; VAT variable in number between species. Non-annulated tail region ornamented with punctations, setae variable in number and location. Differentiated from other 2 stages by 4 CAT, 3 rows of PAT.

Key to Species of *Paradraconema* n. gen.

1. Male amphids elongate loop-shaped or elongate unispiral. Female amphids elongate loop-shaped; or unispiral, spiral never doubled more than $^1/_4$ 2
   Male amphids doubled elongate spiral, $^1/_2$ to $^3/_4$ spiral doubled. Female amphids doubled circular or elongate spiral, $^1/_2$ to $^3/_4$ spiral doubled . . . . *antarcticum* n. sp. (p. 62)
2. Males and females without longitudinally areolated annules on swollen esophageal region . . . 3
   Males and females with longitudinally areolated annules on swollen esophageal region (fig. 75) . 5
3. Males and females with large Ceph Acan-set on posterior $^1/_3$ of rostrum (fig. 88), $^1/_2$ [6] to 4 $^1/_2$ annule widths or 1 to 9 μm anterior to first body annule 4
   Males and females with large Ceph Acan-set about mid-rostrum (fig. 66), 8 to 12 annule widths or 17 to 20 μm anterior to first body annule . . . . *californicum* n. sp. (p. 50)
4. Males with single ventral preanal Corn-set. Female amphids usually elongate loop-shaped, ventral arm sometimes partially converged toward dorsal arm almost forming unispiral; Ceph Acan-set 2 to 4 $^1/_2$ annule widths or 5 to 9 μm anterior to first body annule . . . .
   . . . . *meridionale* (Kreis, 1938) n. comb. (p. 56)
   Males with 1 subventral pair of preanal Corn-set. Female amphids circular unispiral, distal end of ventral arm directed anteriorly; Ceph Acan-set $^1/_2$ to 2 annule widths or 1 to 3 μm anterior to first body annule . . . . *newelli* n. sp. (p. 54)
5. Males without (fig. 62) or with 1 pair of subventral preanal Acan-set (fig. 81), 3 to 5 μm long. Females with 12 to 15 SlAT 6
   Males with single ventral preanal Corn-set (fig. 95), 5 to 7 μm long. Females with 17 to 18 SlAT
   . . . . *hopperi* n. sp. (p. 58)
6. Males with 1 pair of preanal Acan-set. Females with 15 to 20 SvAT 7
   Males without preanal Acan-set. Females with 24 to 30 SvAT . . . .
   . . . . *spinosum* (Southern, 1914) n. comb. (p. 49)
7. Males and females with prominent eyespots (fig. 50). Males T/ABD 4.1 to 5.8, gubernaculum (Gub) 10 to 19 μm long, preanal Acan-set 3 to 5 μm long. Females with 1 fine, obscure, subdorsal pair of setae about 33% to 50% on non-annulated tail region; Ceph Acan-set 2 to 4 annule widths or 4 to 9 μm anterior to first body annule . . . . . *floridense* n. sp. (p. 46)
   Males and females with obscure eyespots. Males T/ABD 3.6 to 3.8, Gub 21 to 22 μm long, preanal Acan-set 3 μm long. Females with 1 subdorsal pair of setae, present or absent, about 66% on non-annulated tail region; Ceph Acan-set 1 $^1/_2$ to 2 annule widths or 2 to 6 μm anterior to first body annule . . . . *singaporense* n. sp. (p. 55)

Paradraconema floridense n. sp.

(Figs. 49 to 55)

Measurements (27 ♀♀): L = 0.8 (0.6-1.1) mm; b = 9.1 (7.2-11.7); c = 8.0 (6.2-11.0); V = 51 (47-54)%; CAT = 24 (20-29) μm; SER (L/W) = 2.4 (2.0-2.7); SER (E/L) = 15 (13-17)%; SS = 13-62 μm; first SlAT = 45 (32-61) μm; last SlAT = 32 (22-44) μm; first SvAT = 40 (29-53) μm; last SvAT = 21 (15-24) μm; No SlAT = 13 (12-14); No SvAT = 17 (15-20); Non-ann Term to Tail Length = 40 (33-49)%; T/ABD = 6.3 (5.2-7.2); Ceph Acan-set = 4 (3-6) μm.

(26 ♂♂): L = 0.8 (0.6-1.1) mm; a = 17.6 (11.9-23.3); b = 10.3 (8.2-12.3); c = 7.3 (6.2-9.0); CAT = 24 (17-29) μm; SER (L/W) = 2.6 (2.1-3.1); SER (E/L) = 15 (14-17)%; SS = 11-58 μm; first SlAT = 42 (32-53) μm; last SlAT = 30 (21-39) μm; first SvAT = 38 (27-49) μm; last SvAT = 20 (15-26) μm; No SlAT = 10 (10-11); No SvAT = 16 (14-16); Non-ann Term to Tail Length = 35 (31-40)%; T/ABD = 5.0 (4.1-5.8); Spic = 48 (33-58) μm; Gub = 15 (10-19) μm; Ceph Acan-set = 4 (2-6) μm; Preanal Acan-set = 4 (3-5) μm.

(Holotype ♂): L = 0.9 mm; a = 17.3; b = 11.3; c = 7.5; CAT = 25 μm; SER (L/W) = 2.8; SER (E/L) = 15%; SS = 21-53 μm; first SlAT = 49 μm; last SlAT = 32 μm; first SvAT = 38 μm; last SvAT = 21 μm; No SlAT = 10; No SvAT = 15; Non-ann Term to Tail Length = 34%; T/ABD = 4.6; Spic = 55 μm; Gub = 16 μm; Ceph Acan-set = 3 μm; Preanal Acan-set = 4 μm.

6. All counts and measurements on right side of body.

Male Holotype.—(Figures in parentheses refer to range within species.) Rostrum with faint subcuticular markings. Amphids large, elongate loop-shaped. CAT typical of genus. One pair of Ceph Acan-set on posterior $1/2$ of rostrum, 3 (2 to 3) annule widths or 7 μm (6 to 8 μm) anterior to first body annule. Prominent eyespots. Annules longitudinally areolated in anterior $1/3$ of body; rest of body annules finer, without ornamentation. First 10 (5 to 10) body annules on swollen esophageal region with distinct subcuticular markings and vacuoles, margins directed anteriorly. Longest SS on swollen esophageal region. Long and short setae intermingled with SlAT. SlAT with 1 (1 to 2) long setae not alternating with tubes, setae equal to or longer than adjacent tubes. SlAT with 5 long and 5 short tubes (3 to 7 long, 3 to 7 short). Caudal glands extend anterior to anus 2.4 (1.8 to 4.4) times ABD. One pair of subventral preanal Acan-set. Three pairs of anal setae, 2 (1 to 2) pairs anterior and 1 (1 to 2) pair posterior to anus, located close together and close to anus, 8 to 9 μm (8 to 10 μm) long; distal ends directed anteriorly. Anal flap short, not crenate. Two (1 to 2) pairs of setae on non-annulated tail region, position measured from last complete tail annule to tail tip; 1 subdorsal pair about 33%, 1 subventral (lateral to subventral) pair just posterior to last complete tail annule (absent on some specimens).

Females.—Similar to males. Amphids usually elongate loop-shaped, can be unispiral. Ceph Acan-set 2 to 4 annule widths or 4 to 9 μm anterior to first body annule. First 5 to 13 annules on swollen esophageal region as in males. Longest SS on swollen esophageal region. Short setae intermingled with SlAT. SlAT with 3 to 8 long and 4 to 10 short tubes. Caudal glands extend anterior to anus 2.5 to 3.6 times ABD. Anal flaps crenated. Two pairs of setae on non-annulated tail region; 1 subdorsal pair 33% to 50%; usually 1 lateral to subventral pair just posterior to last complete tail annule or on last annule.

Second-Stage Larvae.—Not observed.

Third-Stage Larvae.—L = 0.4 to 0.6 mm. Similar to adults. Amphids elongate loop-shaped or elongate unispiral. CAT typical of genus. One pair of Ceph Acan-set 1 to 2 μm long, 4 to 7 annule widths or 8 to 11 μm anterior to first body annule. Eyespots present. First 3 body annules on swollen esophageal region ornamented as in adults, remaining annulation without ornamentation. SlAT typical of genus. Anal flap long, slightly crenate. One subdorsal pair of setae on non-annulated tail region 33% to 50%; some specimens with single dorsal seta about 33%.

Fourth-Stage Larvae.—L = 0.7 mm. Similar to adults. Amphids elongate loop-shaped to almost unispiral, ventral arm converging toward dorsal arm. CAT typical of genus. One pair of Ceph Acan-set 2 to 3 μm long, 4 to 9 annule widths or 10 to 11 μm anterior to first body annule. Prominent eyespots. First 3 to 7 body annules on swollen esophageal region as in adults, remaining annulation without ornamentation. SlAT typical of genus; 7 to 10 VAT. Anal flap long, crenate. Two pairs of setae on non-annulated tail region; 1 subdorsal pair 50% to 66%; 1 lateral to subventral pair just posterior to last complete tail annule or 1 to 2 annules anterior to last annule.

Holotype (♂).—Collected February 20, 1965 by B. E. Hopper. Catalogue No. UCNC 1417, University of California, Davis.

Paratypes.—18 ♀♀, 9 ♂♂. Same data as holotype. Paratypes deposited at University of California, Davis (UCNC 1464 to 1466); USDA Nematode Collection, Beltsville, Maryland; U. S. National Museum of Natural History, Smithsonian Institution, Washington, D. C.; and Laboratoria voor Morfologie en Systematiek, Museum voor Dierkunde, Gent, Belgium.

Larval Stages.—3 third, 23 fourth. Same data as holotype. Deposited in same nematode collections as paratypes.

Type Habitat.—Marine, associated with *Halimeda* sp., a calcareous alga.

Type Locality.—Coral Key, Florida, USA (Miami Beach side).

Distribution.—Port Jackson, Australia. Solano, Colombia. Bear Cut and Coral Key, Florida, USA. Japan: Ibusuki, Kyushu; and Shimoda, Shizuoka-ken. Galeta Beach and Panama City, Panama. Philippine Islands: Little Santa Cruz Island; Matabungkay, Batangas; and Zamboanga, Mindanao. American Samoa, Pago Pago Island, Samoa Islands. Society Islands: Tiarei, Tahiti Island; Uturoa and Raiatea, Raiatea Island.

Diagnosis.—Most closely resembles *P. singaporense* n. sp. differs by prominent eyespot; males differ by longer tails and preanal Acan-set, shorter gubernaculum, and absence of setae about 66% on non-annulated tail region; and females by 1 subdorsal pair of setae 33% to 50% on non-annulated

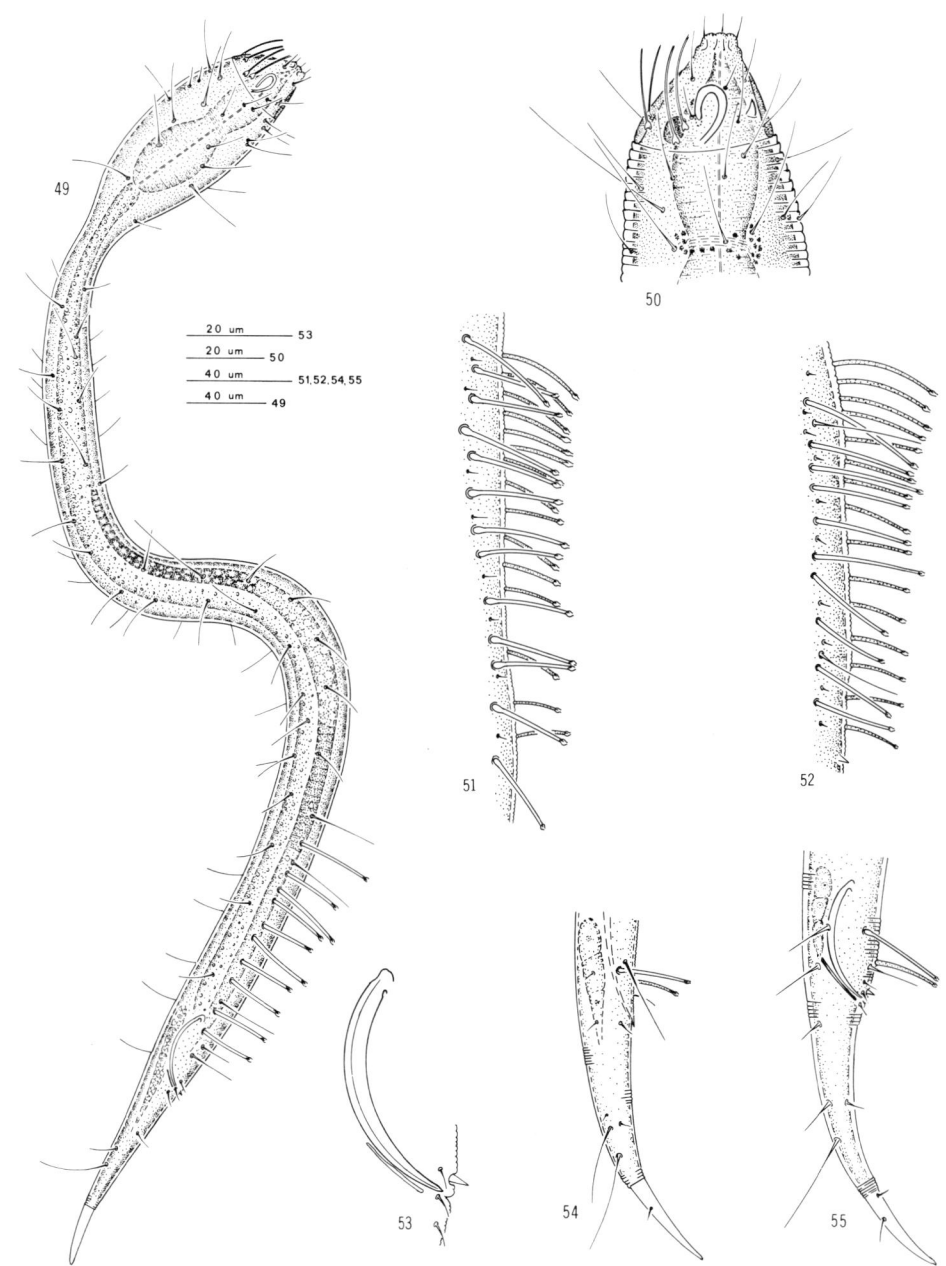

7. Figs. 49-55: *Paradraconema floridense* n. sp. 49) Male, full length[a], SvAT and correct number of CAT not illustrated; 50) Male, head, correct number of CAT not illustrated; 51) Female, PAT; 52) Male, PAT; 53) Male, anal region; 54) Female, tail; 55) Male, tail.

a. Total number of setae on non-annulated tail not illustrated.

tail region, and more anteriorly located Ceph Acan-set on rostrum. Differs from *P. antarcticum* n. sp. by amphids; males elongate loop-shaped; females usually elongate loop-shaped, can be unispiral. Differs from *P. californicum* n. sp., *P. newelli* n. sp., and *P. meridionale* n. comb. by annules longitudinally areolated on swollen esophageal region. Differs from *P. californicum* n. sp. and *P. spinosum* n. comb. by 1 pair of Ceph Acan-set. Differs from *P. spinosum* n. comb. by 1 pair of preanal Acan-set in males, and fewer SvAT in females; from *P. hopperi* n. sp. by males with 1 pair of subventral preanal Acan-set, and fewer SlAT in females.

Third-stage larvae differ from third-stage *P. spinosum* n. comb. and *P. antarcticum* n. sp. by elongate loop-shaped or unispiral amphids, and shorter Ceph Acan-set on rostrum. Fourth-stage most closely resemble fourth-stage *P. newelli* n. sp. differ by larger size, and shorter Ceph Acan-set on rostrum. Differs from fourth-stage *P. spinosum* n. comb., *P. californicum* n. sp., and *P. antarcticum* n. sp. by 1 pair of Ceph Acan-set; from *P. meridionale* n. comb. by fewer VAT; from *P. hopperi* n. sp. by shorter Ceph Acan-set on rostrum.

Paradraconema spinosum (Southern, 1914) n. comb.

(Figs. 56 to 64)

Syn: *Chaetosoma spinosum* Southern, 1914 n. syn.
  Notochaetosoma spinosum (Southern, 1914) Cobb, 1929 n. syn.
  Drepanonema spinosum (Southern, 1914) Cobb, 1933 n. syn.
  Claparediella spinosum (Southern, 1914) Filipjev, 1934 n. syn.
  Draconema spinosum (Southern, 1914) Kreis, 1938 n. syn.

Measurements (9 ♀♀): L = 1.2 (1.0-1.3) mm; b = 9.8 (9.3-10.5); c = 8.6 (7.9-9.7); V = 55 (53-57); CAT = 29 (25-33) μm; SER (L/W) = 2.5 (2.1-2.7); SER (E/L) = 14 (14-15)%; SS = 31-86 μm; first SlAT = 56 (49-65) μm; last SlAT = 32 (31-35) μm; first SvAT = 49 (44-52) μm; last SvAT = 21 (20-24) μm; No SlAT = 15 (14-15); No SvAT = 27 (24-30); Non-ann Term to Tail Length = 36 (35-38)%; T/ABD = 7.6 (6.7-8.8); Ceph Acan-set = 7 (7-9) μm.

(12 ♂♂): L = 1.1 (1.0-1.3) mm; a = 26.6 (22.7-29.0); b = 9.7 (8.3-11.2); c = 7.6 (6.9-8.7); CAT = 30 (27-33) μm; SER (L/W) = 2.7 (2.3-3.2); SER (E/L) = 14 (13-14)%; SS = 24-94 μm; first SlAT = 57 (50-65) μm; last SlAT = 37 (33-45) μm; first SvAT = 49 (47-51) μm; last SvAT = 20 (15-23) μm; No SlAT = 12 (11-13); No SvAT = 23 (20-25); Non-ann Term to Tail Length = 29 (27-33)%; T/ABD = 6.1 (5.7-6.8); Spic = 61 (56-75) μm; Gub = 20 (19-21) μm; Ceph Acan-set = 6 (5-7) μm.

Males Emended.—Rostrum with subcuticular markings. Amphids large, conspicuous; elongate loop-shaped, sometimes ventral arm directed anteriorly to form broken elongate or elongated closed unispiral. CAT typical of genus. Two pairs of Ceph Acan-set; large pair on posterior $^1/_3$ of rostrum, 1 to 4 annule widths or 4 to 8 μm anterior to first body annule; small pair about mid-rostrum, more lateral toward amphid, 2 to 3 μm long. Prominent eyespots. Annules longitudinally areolated in anterior $^1/_3$ of body; remaining annulation finer, without ornamentation. First 8 to 10 annules on swollen esophageal region with distinct subcuticular markings and vacuoles, margins directed anteriorly. Annules on swollen esophageal region with minute anteriorly directed spines (fig. 57). Longest SS on swollen esophageal region. Long and short setae intermingled with SlAT. SlAT with 2 long setae, not alternating with tubes, setae longer than adjacent tubes. SlAT with 3 to 8 long and 4 to 9 short tubes. Caudal glands extend anterior to anus 2.5 to 3.2 times ABD. Preanal Acan-set or Corn-set absent. Four pairs of anal setae, usually 2 pairs anterior and 2 posterior to anus, 14 to 19 μm long. Anal flap short, not crenate. One to 3 pairs of setae on non-annulated tail region, position measured from last complete tail annule to tail tip; 1 lateral to subventral pair about 13%; usually 1 subdorsal pair about 33%; 1 subdorsal pair present or absent, 50% to 66%.

Females Emended.—Similar to males. Amphids circular unispiral, only $^1/_4$ spiral doubled (fig. 58). Large Ceph Acan-set 3 to 7 μm anterior to first body annule; small Ceph Acan-set about mid-rostrum, 2 to 5 μm long. First 10 to 11 annules on swollen esophageal region distinct with subcuticular markings and vacuoles. Two pairs of paravulval setae, 8 to 10 μm long, 1 pair anterior and 1 posterior to vulva. Longest SS on swollen esophageal region. Short setae intermingled with SlAT. SlAT with 6 to 9 long and 6 to 8 short tubes. Anal flap crenate (fig. 59). Two pairs of subdorsal setae on non-annulated tail region; 1 pair about 50%, 1 pair about 66%; some specimens with single dorsal seta about 75%.

Second-Stage Larvae.—L = 0.4 to 0.5 mm. Similar to adults. Amphids unispiral. CAT typical of genus. Ceph Acan-set absent. Eyespots present, but obscure. Annulation without ornamentation. SlAT typical of genus. Usually with single dorsal seta or some specimens with 1 subdorsal pair of setae about 33% on non-annulated tail region.

Third-Stage Larvae.—L = 0.5 to 0.7 mm. Similar to adults. Amphids partially doubled circular spiral, ventral arm against dorsal arm forms double spiral ($^1/_4$ to $^1/_2$ spiral doubled). CAT typical of genus. One pair of Ceph Acan-set on rostrum, 2 to 4 μm long, 4 to 7 annule widths or 7 to 11 μm anterior to first body annule. Eyespots present. First 2 to 3 body annules ornamented as in adults, with anteriorly directed margins; remaining annulation without ornamentation. SlAT typical of genus. Anal flap crenate. One subdorsal pair of setae about 50% on non-annulated tail region.

Fourth-Stage Larvae.—L = 0.7 to 0.9 mm. Similar to adults. Amphids large, circular unispiral. CAT typical of genus. Two pairs of Ceph Acan-set on rostrum; large Acan-set 3 to 5 μm long, 3 to 4 annule widths or 5 to 7 μm anterior to first body annule; smaller Acan-set about mid-rostrum, 1 to 2 μm long, sometimes mistaken for raised cuticular areas on rostrum. Eyespots present. Annules longitudinally areolated in anterior $^1/_3$ of body, remaining annulation without ornamentation. First 3 to 5 body annules on swollen esophageal region ornamented as in adults. SlAT typical of genus; 10 to 11 VAT. Anal flap crenate. One subdorsal pair of setae 50% to 66% on non-annulated tail region, some specimens with single dorsal seta about 66%.

Type Habitat.—Marine, associated with sand and shells at 43 meters.

Type Locality.—Clare Island in Clew Bay, Ireland.

Distribution.—Clare Island in Clew Bay, Ireland; and Marseille, France.

Diagnosis.—Most closely resembles *P. californicum* n. sp. differs by longitudinally areolated annules on swollen esophageal region; absence of preanal Acan-set in males; greater number of SlAT and SvAT in females. Differs from *P. hopperi* n. sp., *P. meridionale* n. comb., *P. newelli* n. sp., *P. floridense* n. sp., and *P. singaporense* n. sp. by 2 pairs of Ceph Acan-set on rostrum. Differs from *P. meridionale* n. comb. and *P. newelli* n. sp. by longitudinally areolated annules on swollen esophageal region. Males differ from *P. hopperi* n. sp., *P. floridense* n. sp., and *P. singaporense* n. sp. by absence of preanal Acan-set or Corn-set. Females differ from *P. hopperi* n. sp. by fewer SlAT; from *P. floridense* n. sp. and *P. singaporense* n. sp. by greater number of SvAT. Differs from *P. antarcticum* n. sp. by amphids; males with elongate loop-shaped amphids, sometimes broken elongate spiral or elongate closed unispiral; females with circular unispiral amphids, $^1/_4$ spiral doubled.

Second-stage larvae differ from second-stage *P. antarcticum* n. sp. by smaller size, and unispiral amphids. Third-stage larvae most closely resemble third-stage *P. antarcticum* n. sp. differ by more posteriorly located Ceph Acan-set on rostrum, and 1 subdorsal pair of setae about 50% on non-annulated tail region. Differ from third-stage *P. floridense* n. sp. by spiral amphids, $^1/_4$ to $^1/_2$ spiral doubled. Fourth-stage larvae most closely resemble fourth-stage *P. californicum* n. sp. differ by more posterior large Ceph Acan-set on rostrum, and 1 subdorsal pair of setae 50% to 66% on non-annulated tail region. Differ from fourth-stage *P. antarcticum* n. sp. by unispiral amphids, smaller size, more posterior large Ceph Acan-set on rostrum, greater number of VAT, and absence of subventral setae on non-annulated tail region. Differ from fourth-stage *P. floridense* n. sp., *P. newelli* n. sp., *P. meridionale* n. comb., and *P. hopperi* n. sp. by 2 pairs of Ceph Acan-set on rostrum.

Type material of this species was not available for study. Southern's 1914 original description was quite adequate. From the distinctive female characters, amphid shape, 2 pairs of Ceph Acan-set on rostrum, and number of SvAT, we were able to identify specimens as belonging to this species.

Paradraconema californicum n. sp.

(Figs. 65 to 71)

Measurements (2 ♀♀): L = 0.9 mm; b = 7.9 (7.2-8.6); c = 8.6; V = 51%; CAT = 28 (28-29) μm; SER (L/W) = 1.8; SER (E/L) = 16 (15-16)%; SS = 15-52 μm; first SlAT = 47 (45-50) μm; last SlAT = 35 (34-36) μm; first SvAT = 48 μm; last SvAT = 26 (25-27) μm; No SlAT = 13 (12-

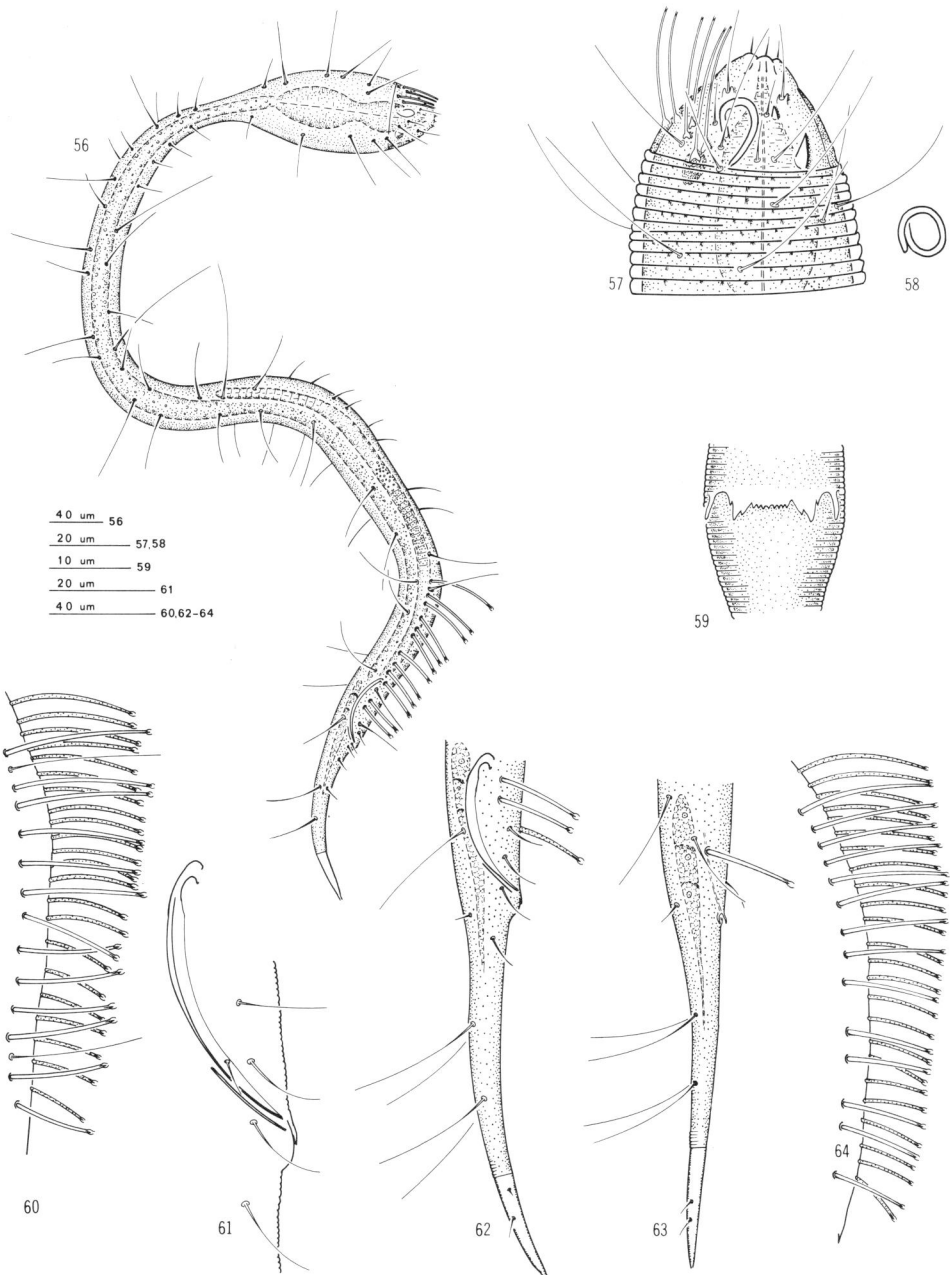

8. Figs. 56–64: *Paradraconema spinosum* (Southern, 1914) n. comb. 56) Male, full length[a], SvAT and correct number of CAT not illustrated; 57) Male, head; 58) Female, amphid; 59) Female, crenate anal flap, ventral view; 60) Male, PAT[b]; 61) Male, anal region; 62) Male, tail; 63) Female, tail; 64) Female, PAT[b].

a. Total number of setae on non-annulated tail not illustrated.
b. Short SS with SlAT not illustrated.

13); No SvAT = 22 (21-22); Non-ann Term to Tail Length = 45 (41-48)%; T/ABD = 4.4 (4.2-4.6); Ceph Acan-set = 5 μm.

(Holotype ♂): L = 0.9 mm; a = 12.8; b = 11.5; c = 7.7; CAT = 27 μm; SER (L/W) = 1.9; SER (E/L) = 15%; SS = 14-54 μm; first SlAT = 44 μm; last SlAT = 36 μm; first SvAT = 44 μm; last SvAT = 25 μm; No SlAT = 10; No SvAT = 18; Non-ann Term to Tail Length = 36%; T/ABD = 4.1; Spic = 53 μm; Gub = 13 μm; Ceph Acan-set = 5 μm; Preanal Acan-set = 4 μm.

Male Holotype.—Rostrum with faint subcuticular markings. Amphids elongate loop-shaped. CAT typical of genus. Two pairs of Ceph Acan-set; large pair about mid-rostrum, 10 annule widths or 19 μm anterior to first body annule; small pair on anterior $1/2$ of rostrum, 2 μm long, more lateral toward amphid than large pair. Obscure eyespots present. Annulation without ornamentation. Longest SS on swollen esophageal region. Long and short setae intermingled with SlAT. SlAT with 3 long setae, not alternating with tubes, setae equal to or slightly longer than adjacent tubes. SlAT with 4 long and 6 short tubes. Caudal glands extend anterior to anus 2.8 times ABD. One pair of subventral preanal Acan-set. Three pairs of anal setae 5 to 9 μm long, 1 pair anterior and 2 posterior to anus. Anal flap absent. Two subventral pairs of setae on non-annulated tail region, 1 pair just posterior to last complete tail annule, 1 pair just posterior to first pair.

Females.—Similar to males. Amphids elongate loop-shaped, ventral arm directed anteriorly toward dorsal arm forming partially closed elongated spiral. Ceph Acan-set with large pair 8 to 12 annule widths or 17 to 19 μm anterior to first body annule; small pair 2 to 3 μm long, on anterior $1/2$ of rostrum. Two pairs of paravulval setae, 6 to 10 μm long, 1 pair anterior and 1 posterior to vulva. Longest SS on swollen esophageal region. Short setae intermingled with SlAT. SlAT with 4 to 5 long and 8 to 9 short tubes. Caudal glands extend anterior to anus 3.1 to 3.7 times ABD. Anal flap faintly crenated. Setae on non-annulated tail region, position measured from last complete tail annule to tail tip; 1 lateral to subventral pair either just posterior to last complete tail annule or about 25%; single dorsal seta about 66%.

Second- and Third-Stage Larvae.—Not observed.

Fourth-Stage Larvae.—L = 0.8 mm. Similar to adults. Amphids large, elongate unispiral. CAT typical of genus. Two pairs of Ceph Acan-set on rostrum; large Acan-set about mid-rostrum, 4 μm long, 12 annule widths or 12 μm anterior to first body annule; small Acan-set on anterior $1/2$ of rostrum, 1 μm long, more lateral toward amphid than large pair. Eyespots present. Annules without ornamentation. SlAT typical of genus; 11 VAT. Anal flap long, crenate. Two pairs of setae on non-annulated tail region; 1 sublateral to subventral pair just posterior to last complete tail annule, 1 sublateral to subdorsal pair about 33%.

Holotype (♂).—Collected April 19, 1961 by Ruth N. Johnson. Catalogue No. UCNC 1418, University of California, Davis.

Paratypes.—2 ♀♀, 1 ♂ (broken). Same data as holotype. Paratypes deposited at University of California, Davis (UCNC 1467, 1469).

Larval Stages.—1 fourth. Same data as holotype. Deposited at University of California, Davis.

Type Habitat.—Marine, associated with holdfasts of *Laminaria digitata*.

Type Locality.—Dillon Beach, Marin County, California, USA.

Distribution.—Same as type locality.

Diagnosis.—Most closely resembles *P. spinosum* n. comb. differs by absence of longitudinally areolated annules on swollen esophageal region; 1 pair of subventral preanal Acan-set in males; females differ by smaller size, fewer number of SlAT and SvAT. Differs from *P. antarcticum* n. sp. by smaller size, greater number of SlAT and SvAT, elongate loop-shaped amphids in males, female amphids elongate loop-shaped with ventral arm directed anteriorly toward dorsal arm forming partially closed elongate unispiral. Differs from *P. floridense* n. sp., *P. newelli* n. sp., *P. singaporense* n. sp., *P. meridionale* n. comb., and *P. hopperi* n. sp. by 2 pairs of Ceph Acan-set on rostrum; from *P. floridense* n. sp., *P. singaporense* n. sp., and *P. hopperi* n. sp. by absence of longitudinally areolated annules on swollen esophageal region; from *P. meridionale* n. comb. and *P. newelli* n. sp. by large Ceph Acan-set about mid-rostrum, 8 to 12 annule widths or 17 to 30 μm anterior to first body annule.

Fourth-stage larvae most closely resemble fourth-stage *P. spinosum* n. comb. differ by large Ceph Acan-set more anterior on rostrum, and absence of setae about 50% to 66% on non-annulated tail region. Differ from fourth-stage *P. antarcticum* n. sp. by smaller size, greater number of VAT, unispiral amphids; and from fourth-stage *P. floridense* n. sp., *P. newelli* n. sp., *P. meridionale* n. comb., and *P. hopperi* n. sp. by 2 pairs of Ceph Acan-set on rostrum.

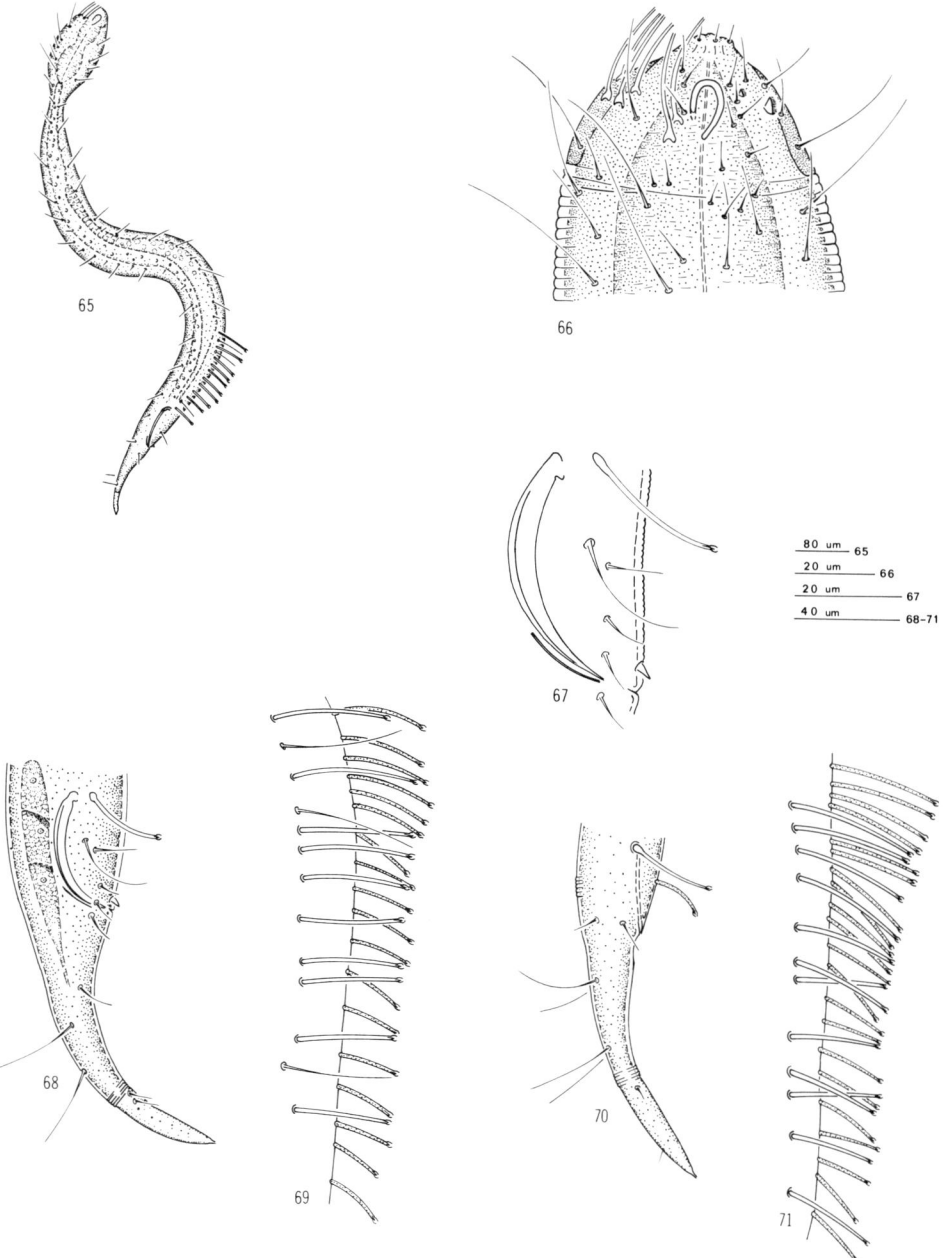

9. Figs. 65-71: *Paradraconema californicum* n. sp. 65) Male, full length[a], SvAT and correct number of CAT not illustrated; 66) Male, head; 67) Male, anal region; 68) Male, tail; 69) Male, PAT[b]; 70) Female, tail; 71) Female, PAT[b].
a. Total number of setae on non-annulated tail region not illustrated.
b. Short SS with SlAT not illustrated.

Paradraconema newelli n. sp.[7]

(Figs. 72 to 73, 76, 79 to 80, 82 to 84)

Measurements (21 ♀♀): L = 0.7 (0.6–0.8) mm; b = 10.0 (8.0–13.3); c = 6.8 (5.3–8.4); V = 52 (47–56)%; CAT = 19 (16–22) μm; SER (L/W) = 1.9 (1.6–2.3); SER (E/L) = 13 (11–15)%; SS = 11–47 μm; first SlAT = 34 (30–39) μm; last SlAT = 25 (21–28) μm; first SvAT = 32 (27–35) μm; last SvAT = 19 (15–22) μm; No SlAT = 12 (10–13); No SvAT = 19 (16–21); Non-ann Term to Tail Length = 31 (28–35)%; T/ABD = 6.3 (5.7–6.6); Ceph Acan-set = 5 (3–7) μm.

(21 ♂♂): L = 0.7 (0.6–0.8) mm; a = 13.8 (12.0–17.0); b = 10.1 (7.5–12.7); c = 7.5 (5.7–10.7); CAT = 21 (17–26) μm; SER (L/W) = 1.8 (1.5–2.0); SER (E/L) = 13 (11–15)%; SS = 10–47 μm; first SlAT = 33 (31–38) μm; last SlAT = 25 (24–28) μm; first SvAT = 31 (24–36) μm; last SvAT = 18 (12–23) μm; No SlAT = 10 (9–10); No SvAT = 17 (13–19); Non-ann Term to Tail Length = 25 (22–27)%; T/ABD = 4.3 (3.5–4.9); Spic = 50 (45–53) μm; Gub = 15 (12–19) μm; Ceph Acan-set = 5 (4–6) μm; Preanal Corn-set = 7 (5–8) μm.

(Holotype ♂): L = 0.7 mm; a = 14.2; b = 8.5; c = 8.5; CAT = 19 μm; SER (L/W) = 1.7; SER (E/L) = 14%; SS = 12–45 μm; first SlAT = 36 μm; last SlAT = 26 μm; first SvAT = 34 μm; last SvAT = 19 μm; No SlAT = 10; No SvAT = 17; Non-ann Term to Tail Length = 27%; T/ABD = 4.3; Spic = 51 μm; Gub = 15 μm; Ceph Acan-set = 6 μm; Preanal Corn-set = 7 μm.

Male Holotype.—(Figures in parentheses refer to range within species.) Rostrum with faint subcuticular markings. Amphids elongate loop-shaped. CAT typical of genus. One pair of Ceph Acan-set on posterior part of rostrum, 1 ($^1/_2$ to 2) annule widths or 1 μm (1 to 3 μm) anterior to first body annule. Obscure eyespots present. Annules in anterior $^1/_3$ of body with distinct subcuticular markings and vacuoles; remaining body annulation finer, without ornamentation. Longest SS on swollen esophageal region. Long and short setae intermingled with SlAT. SlAT with 2 long setae, not alternating with tubes, anterior seta equal in length to adjacent tube and posterior seta longer than adjacent tube. SlAT with 5 (2 to 5) long and 4 (4 to 8) short tubes. Caudal glands extend anterior to anus 2.8 (2.4 to 3.2) times ABD. One pair of subventral preanal Corn-set. Three pairs of anal setae 9 to 12 μm (5 to 12 μm) long, 1 (1 to 2) pair anterior and 2 (1 to 2) pairs posterior to anus, with distal ends directed anteriorly. Anal flap short, not crenated. One subventral pair of setae just posterior to last complete tail annule on non-annulated tail region.

Females.—Similar to males. Amphids circular unispiral. Ceph Acan-set 1 to 3 μm anterior to first body annule. Longest SS on swollen esophageal region. Short setae intermingled with SlAT. SlAT usually with 5 to 7 long and 6 to 8 short tubes. Caudal glands extend anterior to anus 3.6 to 4.4 times ABD. Anal flap short, not crenated. One to 2 pairs of setae on non-annulated tail region, position measured from last complete tail annule to tail tip; 1 subventral pair just posterior to or at level of last complete tail annule; 1 subdorsal pair present or absent, 50% to 66%.

Second- and Third-Stage Larvae.—Not observed.

Fourth-Stage Larvae.—L = 0.5 to 0.6 mm. Similar to adults. Rostrum with subcuticular markings and granules, setae present. Amphids circular unispiral. CAT typical of genus. One pair of Ceph Acan-set on posterior part of rostrum, 3 to 4 μm long, 2 to 3 annule widths or 3 to 6 μm anterior to first body annule. Eyespots present. First 4 to 6 annules on swollen esophageal region with distinct subcuticular markings and vacuoles, remaining annulation without ornamentation. SlAT typical of genus, 7 to 9 VAT. Anal flap not crenated. Two pairs of setae on non-annulated tail region; 1 lateral to sublateral pair just posterior to or 2 to 4 annules anterior to last complete tail annule, 1 subdorsal pair about 66%.

Holotype (♂).—Collected May 27, 1968 by I. M. Newell. Catalogue No. UCNC 1419, University of California, Davis.

Paratypes.—35 ♀♀, 25 ♂♂. Same data as holotype. Paratypes deposited at University of California, Davis (UCNC 1470 to 1472); USDA Nematode Collection, Beltsville, Maryland; U. S. National Museum of Natural History, Smithsonian Institution, Washington, D. C.; Station Marine D'Endoune et Centre D'Oceanographie, Marseille, France; and Laboratoria voor Morfologie en Systematiek, Museum voor Dierkunde, Gent, Belgium.

7. Named in honor of I. M. Newell.

Larval Stages. —9 fourth. Same data as holotype. Deposited at University of California, Davis; USDA Nematode Collection, Beltsville, Maryland; and Laboratoria voor Morfologie en Systematiek, Museum voor Dierkunde, Gent, Belgium.

Type Habitat. —Marine, associated with coral in intertidal zone.

Type Locality. —Twenty miles north of Solano, Colombia.

Distribution. —Solano, Colombia and Isla Taboga, Panama.

Diagnosis. —Most closely resembles *P. meridionale* n. comb. differs by 1 subventral pair of preanal Corn-set in males; females by circular unispiral amphids, and Ceph Acan-set $^{1}/_{2}$ to 2 annule widths or 1 to 3 µm anterior to first body annule. Differs from *P. antarcticum* n. sp., *P. hopperi* n. sp., *P. spinosum* n. comb., *P. floridense* n. sp., and *P. singaporense* n. sp. by absence of longitudinally areolated annules on swollen esophageal region. Differs from *P. hopperi* n. sp. by larger preanal Corn-set in males, fewer SlAT in females; from *P. spinosum* n. comb., *P. californicum* n. sp., and *P. antarcticum* n. sp. by 1 pair of Ceph Acan-set on rostrum.

Fourth-stage larvae most closely resemble fourth-stage *P. hopperi* n. sp. differ by smaller size, shorter and more posteriorly located Ceph Acan-set on rostrum. Differ from fourth-stage larvae of *P. antarcticum* n. sp. and *P. floridense* n. sp. by smaller size, unispiral amphids, and posteriorly located Ceph Acan-set on rostrum. Differ from fourth-stage *P. meridionale* n. comb., and *P. californicum* n. sp. by smaller size, and fewer VAT; and from fourth-stage *P. spinosum* n. comb. by smaller size, presence of setae just posterior to last complete tail annule on non-annulated tail region, fewer VAT.

Paradraconema singaporense n. sp.

(Figs. 74 to 75, 77 to 78, 81, 85 to 86)

Measurements (4 ♀♀): L = 0.9 (0.9–1.0) mm; b = 9.8 (8.3–11.3); c = 8.0 (7.2–9.0); V = 48 (45–51)%; CAT = 25 (22–28) µm; SER (L/W) = 2.5 (2.2–2.8); SER (E/L) = 14 (14–15)%; SS = 16–52 µm; first SlAT = 42 (36–47) µm; last SlAT = 29 (26–32) µm; first SvAT = 43 (40–45) µm; last SvAT = 21 (20–22) µm; No SlAT = 13 (12–14); No SvAT = 17 (16–18); Non-ann Term to Tail Length = 40 (33–48)%; T/ABD = 5.7 (5.2–6.4); Ceph Acan-set = 4 (3–6) µm.

(2 ♂♂): L = 0.9 (0.8–0.9) mm; a = 16.8 (15.3–18.2); b = 10.8 (10.0–11.5); c = 6.7 (5.7–7.7); CAT = 28 µm; SER (L/W) = 2.6 (2.5–2.7); SER (E/L) = 14%; SS = 15–43 µm; first SlAT = 48 (47–49) µm; last SlAT = 32 (31–33) µm; first SvAT = 45 (44–46) µm; last SvAT = 20 (18–23) µm; No SlAT = 10 (9–10); No SvAT = 16; Non-ann Term to Tail Length = 32 (30–33)%; T/ABD = 3.7 (3.6–3.8); Spic = 49 (48–50) µm; Gub = 21 (21–22) µm; Ceph Acan-set = 4 (4–5) µm; Preanal Acan-set = 3 µm.

(Holotype ♂): L = 0.9 mm; a = 15.3; b = 11.5; c = 7.7; CAT = 28 µm; SER (L/W) = 2.7; SER (E/L) = 14%; SS = 15–43 µm; first SlAT = 47 µm; last SlAT = 33 µm; first SvAT = 44 µm; last SvAT = 23 µm; No SlAT = 9; No SvAT = 16; Non-ann Term to Tail Length = 30%; T/ABD = 3.6; Spic = 50 µm; Gub = 22 µm; Ceph Acan-set = 4 µm; Preanal Acan-set = 3 µm.

Male Holotype. —(Figures in parentheses refer to range within species.) Rostrum with faint subcuticular markings. Amphids large, elongate loop-shaped. CAT typical of genus. One pair of Ceph Acan-set on posterior region of rostrum, 3 (1 to 3) annule widths or 5 µm (2 to 5 µm) anterior to first body annule. Obscure eyespots present. Annules longitudinally areolated on anterior $^{1}/_{3}$ of body and posterior to anus; intervening annulation finer, without ornamentation. First 5 (5 to 6) annules on swollen esophageal region with distinct subcuticular markings and vacuoles. Longest SS on swollen esophageal region. Long and short setae intermingled with SlAT. SlAT with 2 long setae, not alternating with tubes, setae about equal to length of adjacent tubes. SlAT with 3 (3 to 4) long and 6 short tubes. Caudal glands extend anterior to anus 2.3 times ABD. One pair of subventral preanal Acan-set. Three pairs of broad-based anal setae 7 to 8 µm long, 1 pair anterior and 2 posterior to anus. Short anal flap. One lateral to subventral pair of setae just posterior to last complete tail annule on non-annulated tail region; with single subdorsal seta about 66% (absent on some specimens) on left side of body, position measured from last complete tail annule to tail tip.

Females. —Similar to males. Ceph Acan-set 1 $^{1}/_{2}$ to 2 annule widths or 2 to 6 µm anterior to first body annule. Annules longitudinally areolated on anterior $^{1}/_{3}$ of body, remaining annulation without ornamentation. First 5 to 7 annules on swollen esophageal region with distinct subcuticular markings and vacuoles. Longest SS on swollen esophageal region. Short setae intermingled with SlAT. SlAT with 4 to 6 long and 6 to 9 short tubes. Caudal glands extend anterior to anus 2.3 to 3.4 times ABD. Anal flap short, not crenated. One to 2 pairs of setae on non-annulated tail region; 1 lateral to subventral pair just posterior to or 3 to 4 annule widths posterior to last complete tail annule; 1 subdorsal pair present or absent, about 66%; some specimens setae not paired and only on one side of body.

Second-, Third- and Fourth-Stage Larvae.—Not observed.

Holotype (♂).—Collected September 27, 1967 by I. M. Newell, T. S. Key, and L. K. Hean. Catalogue No. UCNC 1420, University of California, Davis.

Paratypes.—5 ♀♀, 3 ♂♂. Same data as holotype. Paratypes deposited at University of California, Davis (UCNC 1473 to 1474); and USDA Nematode Collection, Beltsville, Maryland.

Type Habitat.—Marine, associated with *Gracillaria* sp. and other algae.

Type Locality.—Changi Beach, Singapore.

Distribution.—Same as type locality.

Diagnosis.—Most closely resembles *P. floridense* n. sp. differs by obscure eyespots; shorter tail and preanal Acan-set, and longer gubernaculum in males; by more posterior Ceph Acan-set, and presence or absence of 1 subdorsal pair of setae about 66% on non-annulated tail region in females. Differs from *P. californicum* n. sp., *P. meridionale* n. comb., and *P. newelli* n. sp. by longitudinally areolated annules on swollen esophageal region; from *P. antarcticum* n. sp., *P. californicum* n. sp., and *P. spinosum* n. comb. by 1 pair of Ceph Acan-set on rostrum; from *P. hopperi* n. sp. by preanal Acan-set in males, fewer SlAT in females.

Paradraconema meridionale (Kreis, 1938) n. comb.

(Figs. 87 to 91)

Syn: *Draconema meridionalis* Kreis, 1938 n. syn.

Measurements (3 ♀♀): L = 0.9 (0.7–1.1) mm; b = 9.3 (8.5–10.5); c = 8.2 (7.7–8.5); V = 52 (48–54)%; CAT = 24 (22–26) μm; SER (L/W) = 1.8 (1.7–1.9); SER (E/L) = 14 (13–14)%; SS = 15–49 μm; first SlAT = 33 (30–38) μm; last SlAT = 25 (22–27) μm; first SvAT = 29 (27–31) μm; last SvAT = 20 (17–22) μm; No SlAT = 13 (12–13); No SvAT = 18 (18–19); Non-ann Term to Tail Length = 35 (30–40)%; T/ABD = 4.7 (4.1–5.7); Ceph Acan-set = 4 (3–4) μm.

(5 ♂♂): L = 0.9 (0.7–1.1) mm; a = 17.3 (13.8–19.2); b = 10.2 (9.0–11.5); c = 8.4 (6.0–11.0); CAT = 23 (22–25) μm; SER (L/W) = 2.1 (1.8–2.6); SER (E/L) = 13 (13–14)%; SS = 12–49 μm; first SlAT = 38 (37–40) μm; last SlAT = 29 (27–32) μm; first SvAT = 32 (30–34) μm; last SvAT = 22 (21–24) μm; No SlAT = 10; No SvAT = 17 (16–17); Non-ann Term to Tail Length = 17 (13–22)%; T/ABD = 4.1 (3.9–4.4); Spic = 51 (48–54) μm; Gub = 23 (22–25) μm; Ceph Acan-set = 4 (4–6) μm; Preanal Corn-set = 11 (11–13) μm.

Males Emended.—Rostrum with subcuticular markings and granules. Amphids elongate loop-shaped, arms converging together. CAT typical of genus. One pair of Ceph Acan-set on posterior region of rostrum, 4 to 6 annule widths or 4 to 5 μm anterior to first body annule. Prominent eyespots, similar to *P. floridense*. Annules with subcuticular markings and vacuoles on anterior $^1/_3$ of body and posteriorly from between last SlAT and anus; intervening annulation finer, without ornamentation. Longest SS on swollen esophageal region. Long and short setae intermingled with SlAT. SlAT with 2 long setae, not alternating with tubes, setae can be either equal to or longer than adjacent tubes. SlAT with 3 long and 7 short tubes. Caudal glands extend anterior to anus 2.0 to 2.7 times ABD. Single large ventral preanal Corn-set. Three pairs of anal setae, 9 to 17 μm long, 2 pairs anterior and 1 posterior to anus. Anal flap absent. One subventral pair of setae on non-annulated tail region just posterior to last complete tail annule.

Females Emended.—Similar to males. Amphids usually elongate loop-shaped; ventral arm sometimes converging toward dorsal arm, almost forming unispiral. Ceph Acan-set 2 to 5 annule widths or 5 to 9 μm anterior to first body annule. First 7 to 13 annules on swollen esophageal region with subcuticular markings and vacuoles, remaining body annules without ornamentation. Longest SS on swollen esophageal region. Short setae intermingled with SlAT. SlAT usually with 4 long and 9 short tubes. Caudal glands extend anterior to anus 2.4 to 4.4 times ABD. Anal flap short, crenate.

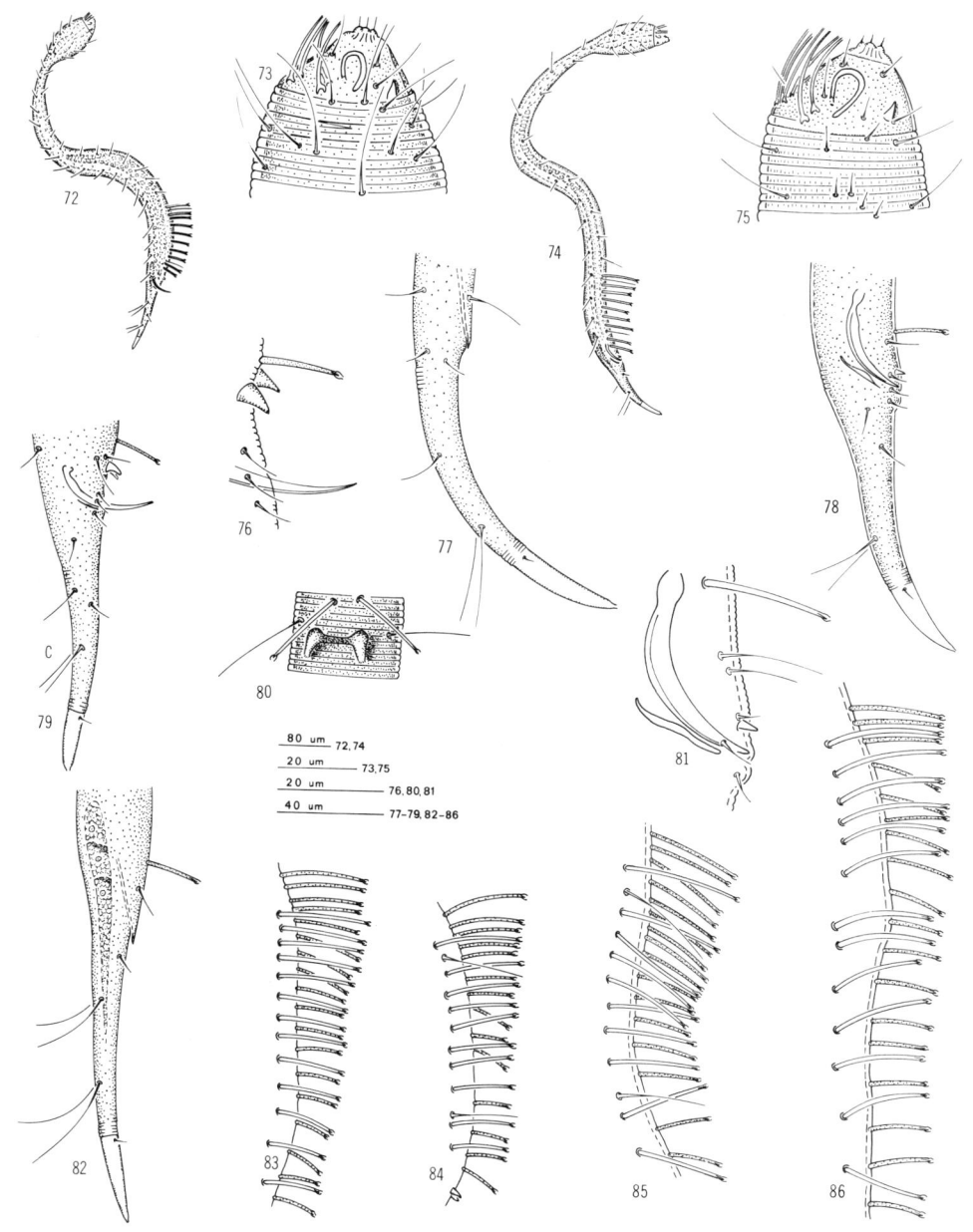

10. Figs. 72-73, 76, 79-80, 82-84: *Paradraconema newelli* n. sp. 72) Male, full length[a], SvAT and correct number of CAT not illustrated; 73) Male, head, correct number of CAT not illustrated; 76) Male, anal region; 79) Male, tail; 80) Male, preanal Corn-set, ventral view; 82) Female, tail; 83) Female, PAT[b]; 84) Male, PAT[b]. Figs. 74-75, 77-78, 81, 85-86: *Paradraconema singaporense* n. sp. 74) Male, full length[a], SvAT and correct number of CAT not illustrated; 75) Male, head; 77) Female, tail; 78) Male, tail; 81) Male, anal region; 85) Male, PAT[b]; 86) Female, PAT[b].

a. Total number of setae on non-annulated tail region not illustrated.
b. Short SS with SlAT not illustrated.

One to 2 pairs of setae on non-annulated tail region, position measured from last complete tail annule to tail tip; 1 subdorsal pair about 66%; 1 lateral to subventral pair present or absent, just posterior to last complete tail annule.

Second- and Third-Stage Larvae.—Not observed.

Fourth-Stage Larvae.—L = 0.8 to 0.9 mm. Similar to adults. Amphids elongate unispiral. CAT typical of genus. One pair of Ceph Acan-set on posterior region of rostrum 3 to 4 μm long, 3 to 5 annule widths or 8 to 10 μm anterior to first body annule. Eyespots present. First 2 annules on swollen esophageal region with faint subcuticular markings and vacuoles, remaining annulation without ornamentation. SIAT typical of genus; 13 VAT. Anal flap not crenated. One subventral pair of setae present or absent, on non-annulated tail region, usually 1 to 4 annules anterior to last complete tail annule, sometimes just posterior to last complete tail annule.

Type Habitat.—Marine.

Locality.—Collected between 1914 and 1916 on Dr. Th. Mortensen Pacific Expedition at various collection sites in the Pacific Ocean: Java Sea Station 110. Banda Sea; Harbor of Amboina, Amboina Island; and along the craggy coast of Kei Island.

Distribution.—Banda Sea: Harbor of Amboina, Amboina Island; and Kei Island. Punta Caldera, Chile. Galapagos Islands: Jensen Island, Santa Cruz Island, and Tower Island. Java Sea.

Diagnosis.—Most closely resembles *P. newelli* n. sp. differs by single preanal Corn-set in males; females by elongate loop-shaped amphids sometimes with arms partially converged almost forms unispiral, Ceph Acan-set 2 to 4 $^1/_2$ annule widths or 5 to 9 μm anterior to first body annule. Differs from *P. antarcticum* n. sp., *P. hopperi* n. sp., *P. spinosum* n. comb., *P. floridense* n. sp., and *P. singaporense* n. sp. by absence of longitudinally areolated annules on swollen esophageal region; from *P. hopperi* n. sp. by longer preanal Corn-set in males, and fewer SIAT in females; from *P. antarcticum* n. sp., *P. californicum* n. sp., and *P. spinosum* n. comb. by 1 pair of Ceph Acan-set on rostrum.

Fourth-stage larvae differ from other known fourth-stage *Paradraconema* by 13 VAT, and 1 pair of setae on non-annulated tail region.

Type material of this species was not available for study; but from the distinctive characters of this species, especially male characters, we were able to identify specimens as belonging to this species.

Paradraconema hopperi n. sp.[8]

(Figs. 92, 94 to 99)

Measurements (16 ♀♀): L = 1.1 (0.9-1.2) mm; b = 9.5 (7.5-10.6); c = 8.5 (6.8-10.0); V = 55 (52-57)%; CAT = 26 (23-28) μm; SER (L/W) = 2.5 (2.1-3.1); SER (E/L) = 14 (12-16)%; SS = 14-70 μm; first SIAT = 47 (40-52) μm; last SIAT = 26 (19-30) μm; first SvAT = 39 (37-42) μm; last SvAT = 18 (13-23) μm; No SIAT = 18 (17-18); No SvAT = 18 (17-19); Non-ann Term to Tail Length = 34 (29-37)%; T/ABD = 7.6 (7.0-8.9); Ceph Acan-set = 6 (5-8) μm.

(11 ♂♂): L = 1.0 (0.8-1.2) mm; a = 19.4 (16.7-22.9); b = 9.8 (8.0-11.8); c = 8.5 (6.7-10.0); CAT = 26 (25-28) μm; SER (L/W) = 2.5 (2.1-2.9); SER (E/L) = 14 (12-15)%; SS = 17-72 μm; first SIAT = 48 (45-52) μm; last SIAT = 27 (25-28) μm; first SvAT = 39 (35-44) μm; last SvAT = 16 (14-19) μm; No SIAT = 13 (12-14); No SvAT = 17 (15-17); Non-ann Term to Tail Length = 30 (25-32)%; T/ABD = 5.0 (4.5-5.5); Spic = 58 (55-63) μm; Gub = 19 (15-21) μm; Ceph Acan-set = 6 (5-8) μm; Preanal Corn-set = 6 (5-7) μm.

(Holotype ♂): L = 1.0 mm; a = 18.8; b = 9.8; c = 9.8; CAT = 25 μm; SER (L/W) = 2.3; SER (E/L) = 12%; SS = 22-65 μm; first SIAT = 48 μm; last SIAT = 26 μm; first SvAT = 40 μm; last SvAT = 19 μm; No SIAT = 14; No SvAT = 17; Non-ann Term to Tail Length = 31%; T/ABD = 4.7; Spic = 56 μm; Gub = 19 μm; Ceph Acan-set = 7 μm; Preanal Corn-set = 6 μm.

Male Holotype.—(Figures in parentheses refer to range within species.) Rostrum with faint subcuticular markings. Amphids elongate loop-shaped. CAT typical of genus. One pair of Ceph Acan-

---

8. Named in honor of B. E. Hopper.

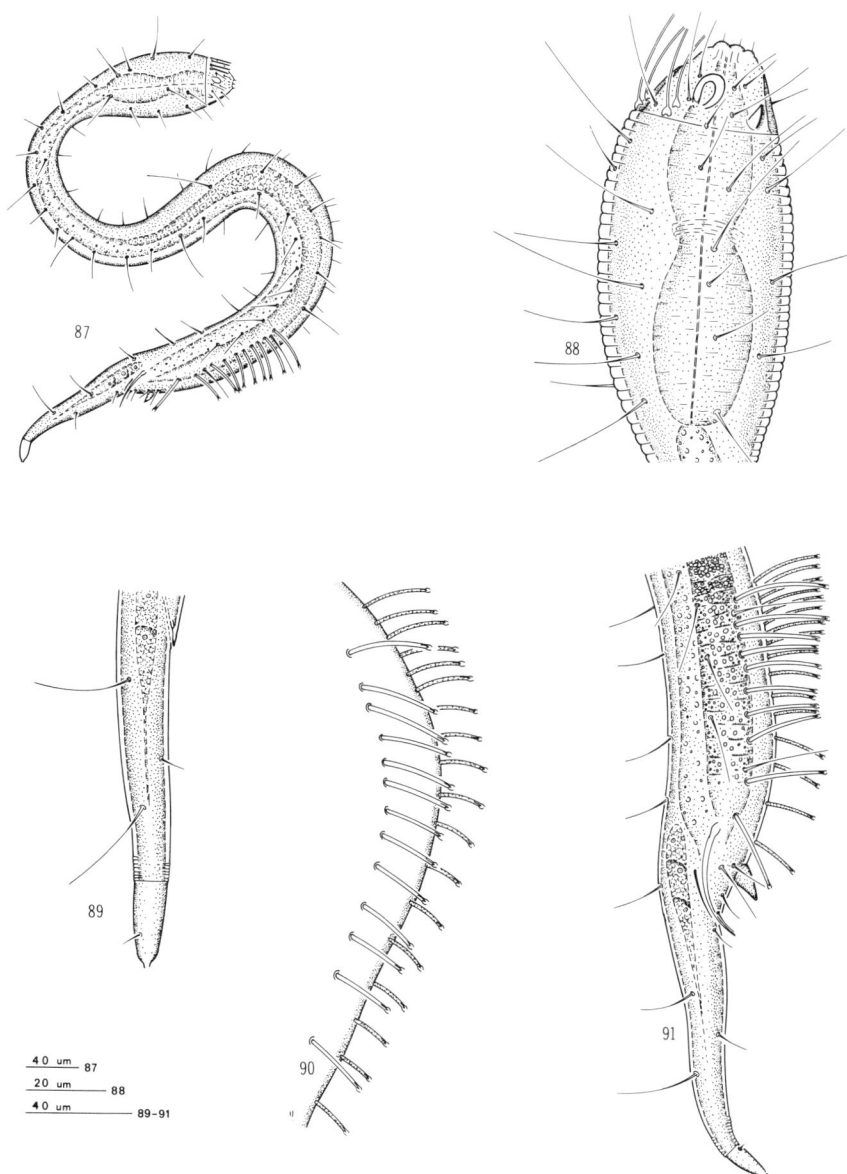

11. Figs. 87–91: *Paradraconema meridionale* (Kreis, 1938) n. comb. 87) Male, full length[a], SvAT and correct number of CAT not illustrated; 88) Male, head, correct number of CAT not illustrated; 89) Female, tail; 90) Female, PAT[b]; 91) Male, posterior region of body, note preanal Corn-set[b].

a. Total number of setae on non-annulated tail region not illustrated.
b. Short SS with SlAT not illustrated.

set on posterior region of rostrum, 1 (½ to 2) annule widths or 3 µm (2 to 3 µm) anterior to first body annule. Prominent eyespots, similar to *P. floridense*. Annules on anterior ⅓ of body and posterior from last SlAT with subcuticular markings and vacuoles; intervening body annulation finer, without ornamentation. Annules on swollen esophageal region longitudinally areolated and with minute anteriorly directed spines. Longest SS on swollen esophageal region. Long and short setae intermingled with SlAT. SlAT with 2 long setae, not alternating with tubes, setae longer than adjacent tubes (some specimens setae shorter than tubes). SlAT with 8 long and 6 short tubes (5 to 8 long and 6 to 9 short tubes). Caudal glands extend anterior to anus 2.7 (2.2 to 3.1) times ABD. Single large ventral preanal Corn-set. Three pairs of broad-based anal setae, 13 to 20 µm long, 2 pairs anterior and 1 posterior to anus. Short anal flap. Two (1 to 2) pairs of setae on non-annulated tail region, position measured from last complete tail annule to tail tip; 1 lateral to subventral pair just posterior to last complete tail annule, 1 subdorsal pair (subventral on some specimens) about 66% (33% to 66%, absent on some specimens).

Females.—Similar to males. Ceph Acan-set 1 to 2 annule widths or 4 to 6 µm anterior to first body annule. Longest SS on swollen esophageal region. Short setae intermingled with SlAT. SlAT with 9 long and 9 short tubes. Caudal glands extend anterior to anus 2.6 to 3.1 times ABD. Anal flap long, not crenated. One to 2 pairs of setae on non-annulated tail region; 1 lateral to subventral pair just posterior to last complete tail annule, usually 1 subdorsal pair 50% to 66%.

Second- and Third-Stage Larvae.—Not observed.

Fourth-Stage Larvae.—L = 0.6 to 0.8 mm. Similar to adults. Amphids elongate loop-shaped or elongate unispiral. CAT typical of genus. One pair of Ceph Acan-set on posterior region of rostrum, 4 to 5 µm long, 3 to 4 annule widths or 6 to 8 µm anterior to first body annule. Eyespots present. Annules on anterior ⅓ of body and posterior from last SlAT to anus with faint subcuticular markings and vacuoles; intervening body annules finer, without ornamentation. First 3 to 6 annules on swollen esophageal region with prominent subcuticular markings and vacuoles. SlAT typical of genus; 9 VAT. Anal flap not crenated. One to 2 pairs of setae on non-annulated tail region; usually 1 subventral pair just posterior to last complete tail annule; 1 subdorsal pair present or absent, 50% to 66%.

Holotype (♂).—Collected February 20, 1965 by B. E. Hopper. Catalogue No. UCNC 1421, University of California, Davis.

Paratypes.—14 ♀♀, 10 ♂♂. Same data as holotype. Deposited at University of California, Davis (UCNC 1475 to 1477); USDA Nematode Collection, Beltsville, Maryland; U. S. National Museum of Natural History, Smithsonian Institution, Washington, D. C.; and Laboratoria voor Morfologie en Systematiek, Museum voor Dierkunde, Gent, Belgium.

Larval Stages.—4 fourth. Same data as holotype. Deposited at University of California, Davis.

Type Habitat.—Marine, associated with *Halimeda* sp., a calcareous alga.

Type Locality.—Coral Key, Florida, USA.

Distribution.—Coral Key and Soldier Key, Florida, USA.

Diagnosis.—Most closely resembles *P. meridionale* n. comb. differs by shorter preanal Corn-set in males, and greater number of SlAT in females. Differs from *P. newelli* n. sp. by single ventral preanal Corn-set in males, greater number of SlAT in females; from *P. floridense* n. sp. and *P. singaporense* n. sp. by single ventral preanal Corn-set in males, and greater number of SlAT in females. Differs from *P. antarcticum* n. sp., *P. californicum* n. sp., and *P. spinosum* n. comb. by 1 pair of Ceph Acan-set on rostrum; from *P. californicum* n. sp., *P. meridionale* n. comb., and *P. newelli* n. sp. by longitudinally areolated annules on swollen esophageal region.

Fourth-stage larvae most closely resemble fourth-stage *P. newelli* n. sp. differ by fewer VAT, larger size, longer and more anterior Ceph Acan-set on the rostrum. Differ from fourth-stage *P. antarcticum* n. sp. by elongate loop-shaped or unispiral amphids, 1 pair of longer and more posterior Ceph Acan-set on rostrum; from *P. spinosum* n. comb. and *P. californicum* n. sp. by 1 pair of Ceph Acan-set on rostrum, and fewer VAT. Differ from fourth-stage *P. meridionale* n. comb. by fewer VAT; from *P. floridense* n. sp. by longer and more posterior Ceph Acan-set on rostrum.

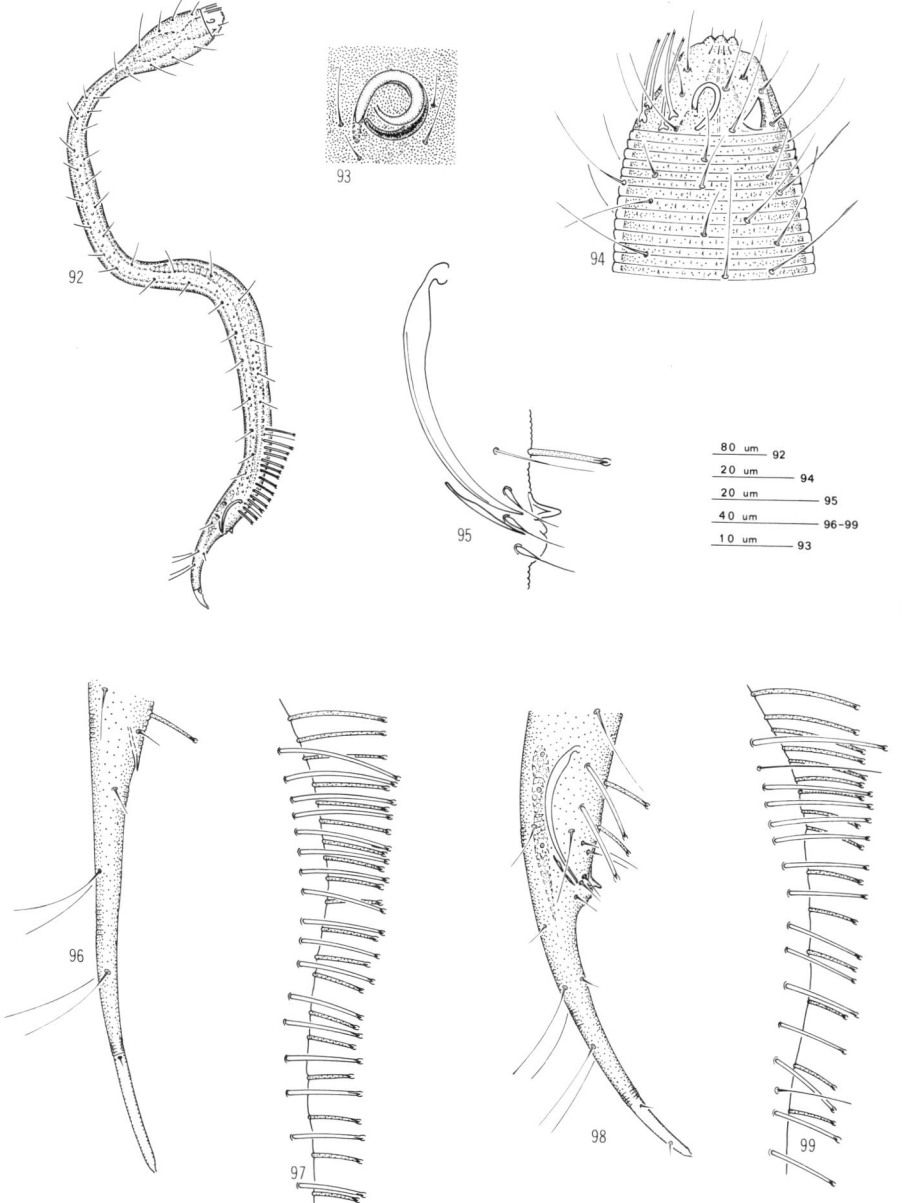

12. Figs. 92, 94–99: *Paradraconema hopperi* n. sp. 92) Male, full length[a], SvAT and correct number of CAT not illustrated; 94) Male, head, correct number of CAT not illustrated; note Ceph Acan-set; 95) Male, anal region; 96) Female, tail; 97) Female, PAT[b]; 98) Male, tail; 99) Male, PAT[b]. Fig. 93: *Paradraconema* sp., female, separation of amphidial tube from groove demonstrating tubular structure of amphid, Cape Agulhas, Republic of South Africa.

a. Total number of setae on non-annulated tail region not illustrated.
b. Short SS with SlAT not illustrated.

Paradraconema antarcticum n. sp.

(Figs. 100 to 107)

Measurements (6 ♀♀): L = 1.4 (1.3-1.5) mm; b = 9.7 (8.1-11.0); c = 8.9 (8.1-9.6); V = 56 (55-57)%; CAT = 33 (30-35) μm; SER (L/W) = 2.8 (2.6-3.3); SER (E/L) = 15 (13-15)%; SS = 16-89 μm; first SlAT = 64 (60-71) μm; last SlAT = 45 (43-47) μm; first SvAT = 53 (52-54) μm; last SvAT = 29 (24-32) μm; No SlAT = 10 (9-10); No SvAT = 16 (16-17); Non-ann Term to Tail Length = 41 (39-43)%; T/ABD = 7.9 (7.4-8.8); Ceph Acan-set = 5 (4-5) μm.

(4 ♂♂): L = 1.5 (1.5-1.6) mm; a = 24.4 (21.1-28.1); b = 10.5 (9.5-11.1); c = 9.2 (8.1-9.8); CAT = 33 (31-35) μm; SER (L/W) = 2.9 (2.7-3.0); SER (E/L) = 14 (13-14)%; SS = 19-91 μm; first SlAT = 64 (62-66) μm; last SlAT = 48 (48-49) μm; first SvAT = 55 (53-59) μm; last SvAT = 31 (29-34) μm; No SlAT = 7 (6-7); No SvAT = 15; Non-ann Term to Tail Length = 39 (36-41)%; T/ABD = 6.8 (6.6-6.9); Spic = 64 (63-66) μm; Gub = 14 (13-16) μm; Ceph Acan-set = 5 (4-6) μm.

(Holotype ♂): L = 1.5 mm; a = 23.8; b = 10.9; c = 9.5; CAT = 35 μm; SER (L/W) = 2.8; SER (E/L) = 13%; SS = 26-91 μm; first SlAT = 62 μm; last SlAT = 49 μm; first SvAT = 53 μm; last SvAT = 29 μm; No SlAT = 7; No SvAT = 15; Non-ann Term to Tail Length = 41%; T/ABD = 6.9; Spic = 63 μm; Gub = 14 μm; Ceph Acan-set = 5 μm.

Male Holotype.—(Figures in parentheses refer to range within species.) Rostrum with faint subcuticular markings and granules. Amphids large, elongate doubled spiral, 1/2 (1/2 to 3/4) spiral doubled. CAT typical of genus. Two pairs of Ceph Acan-set on rostrum; large pair of Ceph Acan-set about mid-rostrum, 6 (5 to 6) annule widths or 12 μm (12 to 15 μm) anterior to first body annule (some specimens distal end may be slightly bifurcate); smaller pair of Ceph Acan-set on anterior 1/2 of rostrum, 2 μm (1 to 2 μm) long, more lateral toward amphid than large pair. Prominent eyespots, similar to *P. floridense*. Annules on anterior 1/3 of body with subcuticular markings and vacuoles, remaining annulation without ornamentation. Annules of swollen esophageal region with faint longitudinal areolation (absent on some specimens). Longest SS on swollen esophageal region. Long and short setae intermingled with SlAT. SlAT with 1 (1 to 3) long setae, not alternating with tubes, setae equal in length to adjacent tubes (some specimens setae either equal to or longer than adjacent tubes). SlAT with 4 long and 3 short tubes (4 long and 2 to 3 short tubes). Caudal glands extend anterior to anus 3.4 (2.3 to 3.4) times ABD. Preanal Acan-set and Corn-set absent. Anal flap short, faintly crenated. Two pairs of anal setae, 12 to 13 μm (10 to 16 μm) long, 1 pair anterior and 1 posterior to anus (some specimens both pairs anterior to anus). Three (1 to 3) pairs of setae on non-annulated tail region, position measured from last complete tail annule to tail tip; 1 subdorsal (subdorsal to lateral) pair (absent on some specimens) about 50%; 2 (1 to 2) pairs subventral (subventral to lateral), 1 pair just posterior to last complete tail annule, 1 pair just posterior to first pair (absent on some specimens). Single fine dorsal seta about 33% on non-annulated tail region (some specimens either with single seta, or 1 subdorsal pair, or setae may be absent).

Females.—Similar to males. Amphids large, doubled elongate or circular spiral, 1/2 to 3/4 spiral doubled. Ceph Acan-set large pair 5 to 9 annule widths or 12 to 17 μm anterior to first body annule. Annules longitudinally areolated on anterior 1/3 of body, remaining annulation without ornamentation. Annules on swollen esophageal region with subcuticular markings and vacuoles. Vulva encircled with minute spine-like projections similar to *Draconema*. Longest SS on swollen esophageal region. Short setae alternate with SlAT. SlAT with 5 to 6 long and 4 to 5 short tubes. Caudal glands extend anterior to anus 2.9 to 3.8 times ABD. Anal flap long, faintly crenated. One to 3 pairs of setae on non-annulated tail region; 1 lateral to subventral pair just posterior to last complete tail annule; usually 2 subdorsal pairs, 1 or both pairs present or absent, 1 about 33%, and 1 about 66%.

Second-Stage Larvae.—L = 0.8 mm. Similar to adults. Rostrum with subcuticular markings, setae present. Amphids doubled circular spiral, 1/4 to 1/2 spiral doubled. CAT typical of genus. Ceph Acan-set absent. Eyespots absent. Annulation without ornamentation. SlAT typical of genus. Anal flap not crenated. Single long dorsal seta about 33% on non-annulated tail terminus.

Third-Stage Larvae.—L = 0.6 to 1.0 mm. Similar to adults. Rostrum with subcuticular markings and granules, setae present. Amphids doubled circular spiral, 1/4 to 1/2 spiral doubled. CAT typical of genus. One pair of Ceph Acan-set about mid-rostrum, 2 to 4 μm long, 7 to 12 annule widths or 11 to 18 μm anterior to first body annule. One specimen with small pair of Ceph Acan-set on anterior $^1/_3$ of rostrum, 2 μm long, more lateral toward amphid than large pair. Eyespots absent. First few annules on swollen esophageal region with faint subcuticular markings and vacuoles, remaining annulation without ornamentation. SlAT typical of genus. Anal flap not crenated. Two pairs of setae on non-annulated tail region; 1 short lateral to subventral pair just posterior to last complete tail annule; 1 long subdorsal pair 50% to 66% about $^1/_2$ length of long setae on annulated tail region; some specimens with single fine dorsal seta about 33%.

Fourth-Stage Larvae.—L = 0.9 to 1.2 mm. Similar to adults. Amphids large, doubled circular spiral, $^1/_2$ to $^3/_4$ spiral doubled. CAT typical of genus. Two pairs of Ceph Acan-set on rostrum; large pair about mid-rostrum, 2 to 4 μm long, 6 to 9 annule widths or 13 to 15 μm anterior to first body annule; small pair on anterior $^1/_3$ of rostrum, 1 to 2 μm long, more lateral toward amphid than large pair. Eyespots present. Annules on anterior $^1/_3$ of body with faint subcuticular markings and vacuoles, remaining annulation without ornamentation. SlAT typical of genus; 8 to 9 VAT. Anal flap crenated. One to 3 pairs of setae on non-annulated tail region; 1 lateral to subventral pair just posterior to last complete tail annule; 1 subdorsal pair 50% to 75%; some specimens with 1 lateral to subventral pair about 75%; usually single fine dorsal seta about 66%.

Holotype (♂).—Collected January 16, 1970 by R. W. Timm and D. R. Viglierchio. U. S. National Museum of Natural History, Smithsonian Institution, Washington, D. C., No. 52005.

Paratypes.—3 ♀♀, 2 ♂♂. Same data as holotype. Two females and 1 male deposited at U. S. National Museum of Natural History, Smithsonian Institution, Washington, D. C.; and 1 female and 1 male deposited at University of California, Davis (UCNC 1488 to 1489).

Larval Stages.—1 third, 2 fourth. Same data as holotype. Deposited in same nematode collections as paratypes.

Type Habitat.—Marine, at 540 meters.

Type Locality.—Scott Base, Antarctica.

Distribution.—Cape Royds, McMurdo Sound, and Scott Base, Antarctica.

Diagnosis.—Most closely resembles *P. spinosum* n. comb. differs by doubled elongate or circular spiral amphids, $^1/_2$ to $^3/_4$ spiral doubled; larger size; and fewer SlAT. Differs from other known *Paradraconema* by larger size; and doubled elongate or circular spiral amphids, $^1/_2$ to $^3/_4$ spiral doubled. Differs from *P. californicum* n. sp., *P. meridionale* n. comb., and *P. newelli* n. sp. by longitudinally areolated annules on swollen esophageal region; from *P. californicum* n. sp. by larger size, fewer SlAT and SvAT. Differs from *P. hopperi* n. sp., *P. meridionale* n. comb., *P. newelli* n. sp., *P. floridense* n. sp., and *P. singaporense* n. sp. by 2 pairs of Ceph Acan-set on rostrum, and larger size.

Second-stage larvae differ from second-stage *P. spinosum* n. comb. by larger size; and doubled circular spiral amphids, $^1/_4$ to $^1/_2$ spiral doubled. Third-stage larvae most closely resemble third-stage *P. spinosum* n. comb. differ by 2 pairs of setae on non-annulated tail region, and more anterior Ceph Acan-set on rostrum. Differ from third-stage *P. floridense* n. sp. by doubled circular spiral amphids, $^1/_4$ to $^1/_2$ spiral doubled; larger size; and 2 pairs of setae on non-annulated tail region. Fourth-stage larvae most closely resemble fourth-stage *P. spinosum* n. comb. differ by doubled circular spiral amphids, $^1/_2$ to $^3/_4$ spiral doubled; larger size; and fewer VAT. Differ from other known fourth-stage *Paradraconema* by doubled circular spiral amphids, $^1/_2$ to $^3/_4$ spiral doubled; and larger size. Differ from *P. californicum* n. sp. by larger size and fewer VAT; from *P. floridense* n. sp., *P. newelli* n. sp., *P. meridionale* n. comb., and *P. hopperi* n. sp. by 2 pairs of Ceph Acan-set on rostrum, and larger size.

Genus *Dracograllus*[9] n. gen.

Diagnosis: Draconematinae. Nematodes 0.4 to 0.9 mm long. Body shape as in *Draconema;* except mid-body of male almost as swollen as female. Most relaxed specimens assume tight "S" shape. Swollen esophageal region averaging 22% of total body length, generally longer than in *Draconema* and *Paradraconema*. Rostrum ornamented with faint subcuticular markings, setae present. Amphids large, conspicuous, slightly dorsal on rostrum. Amphids generally similar in both sexes; usually

---

9. *Dracograllus*, L. f., dragon on stilts.

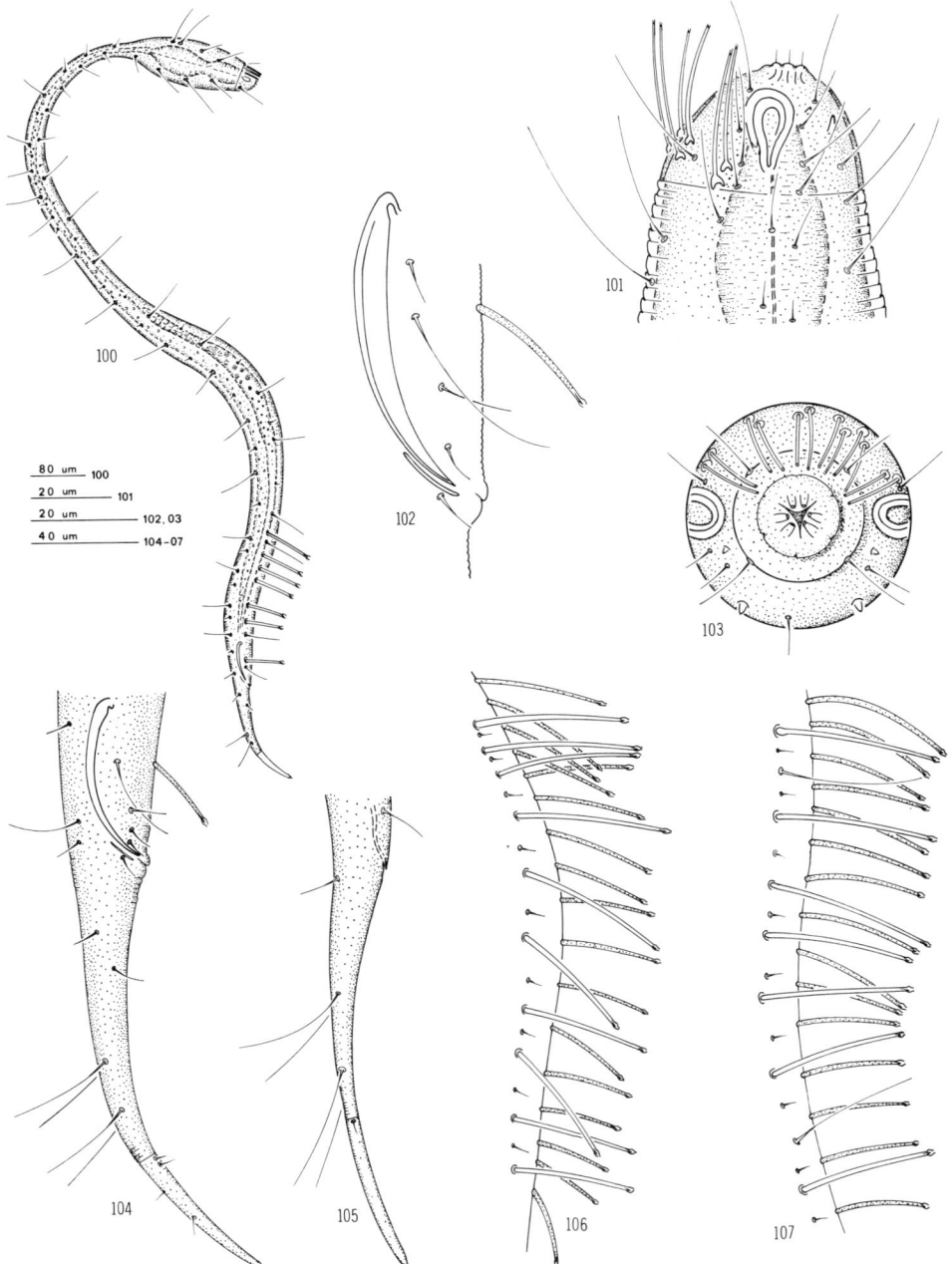

13. Figs. 100–107: *Paradraconema antarcticum* n. sp. 100) Male, full length[a], SvAT and correct number of CAT not illustrated; 101) Male, head; 102) Male, anal region; 103) Male, face view; 104) Male, tail; 105) Female, tail; 106) Female, PAT; 107) Male, PAT.

 a. Total number of setae on non-annulated tail region not illustrated.

elongate loop-shaped, but can be unispiral; ventral arm usually longer than dorsal arm. Some specimens amphidial groove of ventral arm appears to form spiral, but amphid actually elongate loop-shaped. Eight to 15 CAT (figs. 124, 137) on rostrum, paired or unpaired; either posterior, adjacent to or anterior to amphids. Most species without sublateral Ceph Acan-set on rostrum. Body annules with or without subcuticular markings and vacuoles. Sometimes annules ornamented with transverse rows of dot-like punctations; or faint annular ridges with spine-like projections, may appear as 2 rows of fine punctations, similar to fig. 166. Annules sometimes with minute spines, distal ends directed either anteriorly or posteriorly; and sometimes with longitudinally areolated annules. Some species with SS borne on conspicuous, raised, cuticular pedicels (pedicel-setae, fig. 110), or cup-shaped cuticular collars (figs. 118, 122). With or without stout, ventral or subventral setae just anterior to first SvAT. Some species with short, open-ended setae in ventro-sublateral row on annulated tail region. Most species with simple gubernaculum, sometimes with laterally enlarged corpus. Paravulval setae present or absent. PAT paired or unpaired. SlAT with long and short tubes usually not alternating, differences in length conspicuous or inconspicuous. Number of PAT variable between species, some species with SlAT posterior to anus. Preanal Acan-set and Corn-set absent in males. Males with or without uniformly tapered anal setae. Short anal flap present or absent. Setae on non-annulated tail region variable in number and location.

Type Species: *Dracograllus cobbi* n. sp.

Some *Dracograllus* species differ from other known genera in Draconematidae by pedicel-setae (PS). Differs from *Draconema* Cobb, 1913 by absence of prominently enlarged annules on swollen esophageal region, some species differ by SlAT posterior to anus. Differs from *Paradraconema* n. gen. by absence of eyespots; absence in most species of sublateral Ceph Acan-set on rostrum; absence of preanal Acan-set and Corn-set in males; some species differ by SlAT posterior to anus. Differs from *Dracotoranema* n. gen. by absence of slender, alternating, conspicuously long and short tubes in rows of SlAT; and absence of Corn-set in males.

Larval Stages: No first-stage larvae available in *Dracograllus*. Second-stage larvae 0.2 to 0.3 mm long. Similar to adults. Relaxed body either nearly straight or arched ventrally and dorsally. Esophageal region swollen, rest of body nearly cylindrical. Swollen esophageal region slightly longer than in second-stage *Draconema*. Rostrum without ornamentation, setae present. Amphids large, conspicuous, circular unispiral. Single dorsal CAT on rostrum. Six rows of SS on swollen esophageal region, 4 sublateral, 1 ventral and 1 dorsal; 4 sublateral rows on remainder of body. PS present or absent. Two pairs of SlAT in 2 longitudinal rows. Anal flap absent. Non-annulated tail region finely ornamented with punctations, setae present. Differentiated from other 2 stages by 1 CAT, 2 pairs of SlAT in 2 rows.

Third-stage larvae 0.3 to 0.4 mm long. Esophageal region swollen, remainder of body nearly cylindrical. Swollen esophageal region slightly longer than in third-stage *Draconema*. Amphids large, conspicuous, circular unispiral. Three CAT, 2 subdorsal and 1 dorsal, in 2 transverse rows on rostrum. Eight rows of SS on swollen esophageal region, 4 sublateral, 2 subdorsal and 2 subventral. Five rows of SS on mid-body region, 4 sublateral and 1 dorsal; 4 sublateral rows on annulated tail region. Some rows with alternating long and short setae. PS present or absent. SlAT in 2 longitudinal rows, paired or unpaired, tubes variable in number between species. Some species with SlAT posterior to anus. Non-annulated tail region setae variable in number and position. Differentiated from other 2 stages by 3 CAT, 5 or more SlAT in 2 rows.

Fourth-stage larvae 0.4 to 0.7 mm long. Similar to adults. Esophageal region swollen, young females swollen at mid-body. Swollen esophageal region usually longer than in fourth-stage *Draconema*. Amphids large, conspicuous; elongate loop-shaped, or unispiral; ventral arm usually longer than dorsal arm. Four to 6 CAT, in 2 transverse rows on rostrum. Most species without Ceph Acan-set on rostrum. Eight rows of SS on swollen esophageal region, 4 sublateral, 2 subdorsal and 2 subventral. Seven rows of SS on mid-body region, 4 sublateral, 2 subdorsal and 1 ventral; 4 sublateral rows on annulated tail region. Some rows with alternating long and short setae. PS present or absent. Three longitudinal rows of PAT, 2 sublateral and 1 ventral; tubes variable in number between species. SlAT usually paired, some species with SlAT posterior to anus. Non-annulated tail region setae variable in number and location. Differentiated from other 2 stages by 4 to 6 CAT, 3 rows of PAT.

Key to Species of *Dracograllus* n. gen.

1. Twelve to 15 CAT on rostrum (fig. 137) *2*
   Eight CAT on rostrum (fig. 124) *4*
2. Without sublateral Ceph Acan-set on rostrum *3*
   With 1 pair of sublateral Ceph Acan-set on rostrum, about mid-rostrum (fig. 136) . . .
   . . . . *stekhoveni* n. sp. (p. 81)
3. Males with 7 to 8^{10} short, stout setae in subventral rows just anterior to first SvAT, ends directed posteriorly (fig. 131); spicules 39 μm long. Females with 24 SlAT, with 1 tube adjacent to anus and 2 tubes posterior to anus. Males and females length of swollen esophagus to total body length 22% . . . . *gerlachi* n. sp. (p. 79)
   Males with 3 to 4 short, stout setae in subventral rows just anterior to first SvAT, spicules 71 μm long. Females with 21 SlAT, with 3 tubes posterior to anus. Males and females length of swollen esophagus to total body length 13 to 14% . . . .
   . . . . *falcatus* (Irwin-Smith, 1918) n. comb. (p. 87)
4. All CAT adjacent to or posterior to amphids *5*
   All CAT anterior to amphids; with 1 SlAT on non-annulated tail region . . . .
   . . . . *eira* (Inglis, 1968) n. comb. (p. 87)
5. Some somatic setae borne on conspicuous, raised, cuticular pedicels (pedicel-setae; figs. 110, 127); or some setae with distinct, cup-shaped, cuticular collars (figs. 118, 122) *6*
   Somatic setae without pedicel-setae or cup-shaped cuticular collars (fig. 143) *11*
6. Some somatic setae with cuticular collars, collars less than 1 μm long; pedicel-setae absent . *7*
   Some somatic setae with pedicel-setae, pedicels 1 to 8 μm long; cuticular collars absent . . *8*
7. Annules ornamented with a few scattered, minute spines . . . . . *filipjevi* n. sp. (p. 72)
   Annules with faint annular ridges with spine-like projections, sometimes appearing as 2 rows of fine punctations . . . . *timmi* n. sp. (p. 74)
8. Males with 5 to 9 SlAT. Females with 6 to 12 SlAT, all SlAT anterior to anus; 9 to 14 SvAT . *9*
   Males with 12 to 14 SlAT. Females with 15 SlAT, with 1 SlAT posterior to anus; 16 SvAT .
   . . . . *mawsoni* n. sp. (p. 69)
9. Males with 5 to 7 SlAT. Females with 9 to 13 SvAT; caudal glands extend anterior to anus 2.1 to 2.9 times ABD; V at 44% to 52% *10*
   Males with 9 SlAT. Females with 14 SvAT; caudal glands extend anterior to anus 1.1 times ABD; V at 40% . . . . *cobbi* n. sp. (p. 66)
10. Males and females with pedicel-setae in ventro-sublateral rows just anterior to SlAT. Male spicules 45 to 53 μm long. Females with 6 to 8 SlAT . . . . *demani* n. sp. (p. 70)
    Males and females without pedicel-setae in ventro-sublateral rows just anterior to SlAT. Male spicules 36 μm long. Females with 12 SlAT. . . . . *kreisi* n. sp. (p. 78)
11. Males with 6 long somatic setae intermingled with SlAT. Female amphids elongate loop-shaped; with 1 SlAT posterior to anus *12*
    Males with 11 long somatic setae intermingled with SlAT. Female amphids unispiral; with 2 SlAT posterior to anus . . . . *solidus* (Gerlach, 1952) n. comb. (p. 88)
12. Males with 17 SlAT, and 13 SvAT. Females with 14 SlAT, and 12 SvAT . . . . . .
    . . . . *wieseri* n. sp. (p. 84)
    Females with 9 to 10 SlAT, and 8 SvAT . . . . *chitwoodi* n. sp. (p. 76)

Dracograllus cobbi[11] n. sp.

(Figs. 108 to 110)

Measurements (1 ♀): L = 0.5 mm; b = 7.1; c = 7.1; V = 40%; CAT = 21 μm; No CAT = 8; SER (L/W) = 2.0; SER (E/L) = 16%; SS = 12–29 μm; first SlAT = 33 μm; last SlAT = 27 μm; first SvAT = 34 μm; last SvAT = 18 μm; No SlAT = 8; No SvAT = 14; Non-ann Term to Tail Length = 53%; T/ABD = 4.0.

10. All counts and measurements on right side of body.
11. Named in honor of N. A. Cobb.

(Holotype ♂): L = 0.5 mm; a = 11.4; b = 6.3; c = 6.3; CAT = 22 μm; No CAT = 8; SER (L/W) = 2.1; SER (E/L) = 20%; SS = 14–29 μm; first SlAT = 34 μm; last SlAT = 27 μm; first SvAT = 35 μm; last SvAT = 19 μm; No SlAT = 9; No SvAT = 12; Non-ann Term to Tail Length = 44%; T/ABD = 3.7; Spic = 51 μm; Gub = 15 μm.

Male Holotype.—Amphids elongate loop-shaped. Eight CAT, 2 sublateral and 2 subdorsal pairs in 2 transverse rows on rostrum, anterior row adjacent to and posterior row posterior to amphids. Annulation without ornamentation. Longest SS on mid-body region. PS[12] in all 8 rows on swollen esophageal region, pedicels 3 to 6 μm long. Two PS, 5 to 7 μm long, in dorso-sublateral rows just anterior to anterior end of testis. Three PS on right and 4 on left side of body in ventro-sublateral rows just anterior to first SlAT, pedicels 6 to 8 μm long. Five PS on right and 1 on left side of body in subventral rows just anterior to first SvAT, pedicels 3 to 4 μm long. One long pair of PS in dorso-sublateral rows on tail just anterior to last complete tail annule, pedicels 2 μm long. One pair of short, stubby, open-ended setae in ventro-sublateral row on tail just anterior to last complete tail annule (fig. 108). Without stout setae in subventral row just anterior to first SvAT. Long and short setae intermingled with SlAT. Six long setae with SlAT, not alternating with tubes, setae either equal to or shorter than adjacent tubes. Nine SlAT on right and 7 on left side of body. Twelve SvAT on right and 16 on left side of body. SlAT with 5 long and 4 short tubes, not alternating. All SlAT anterior to anus. Caudal glands extend anterior to anus 1.6 times ABD. Two pairs of anal setae, 9 to 13 μm long, 1 pair anterior and 1 posterior to anus. Anal flap absent. Four pairs of setae on non-annulated tail region, position measured from last complete tail annule to tail tip; 1 long latero-subventral pair about 3 annule widths posterior to last complete tail annule about equal in length to long setae on annulated tail region; 3 short subdorsal pairs, 1 pair 2 annule widths posterior to last complete tail annule, 1 about 25%, and 1 about 50%.

Female.—Similar to male. One pair of subventral paravulval setae anterior to vulva, 6 to 7 μm long. Longest SS on swollen esophageal region. PS in all 8 rows on swollen esophageal region, pedicels 3 to 7 μm long. Two PS on right and 3 on left side of body in dorso-sublateral rows just anterior to distal end of anterior ovary, pedicels 5 to 6 μm long. Four PS in ventro-sublateral rows between vulva and first SlAT, pedicels 6 to 7 μm long. Four PS on right and 2 on left side of body in subventral rows, 1 PS on each side just anterior to vulva, remaining PS between vulva and first SvAT, pedicels 3 to 4 μm long. One long pair of PS in dorso-sublateral rows on tail just anterior to last complete tail annule, pedicels 2 μm long. Short setae intermingled with SlAT. Seven long and 1 short tube with SlAT. All SlAT anterior to anus. Caudal glands extend anterior to anus 1.1 times ABD. Short anal flap present. Three pairs of setae on non-annulated tail region; 1 subdorso-lateral pair about 13%; 2 lateral pairs about 33% and 50%.

Second-, Third-, and Fourth-Stage Larvae.—Not observed.

Holotype (♂).—Collected February 20, 1965 by B. E. Hopper. Catalogue No. UCNC 1422, University of California, Davis.

Paratype.—1 ♀. Same data as holotype. Deposited at University of California, Davis (UCNC 1492).

Type Habitat.—Marine, associated with *Halimeda* sp., a calcareous alga.

Type Locality.—Coral Key, Florida, USA.

Distribution.—Same as type locality.

Diagnosis.—Most closely resembles *D. demani* n. sp. and *D. kreisi* n. sp. differs by greater number of SlAT in males; greater number of SvAT, shorter caudal glands, more anterior position of vulva in females. Differs from *D. mawsoni* n. sp. by fewer SlAT in males; and absence of SlAT posterior to anus in females. Differs from *D. gerlachi* n. sp., *D. stekhoveni* n. sp., and *D. falcatus* n. comb. by 8 CAT on rostrum; and from remaining known species of *Dracograllus* by PS.

12. All counts and measurements on right side of body unless indicated.

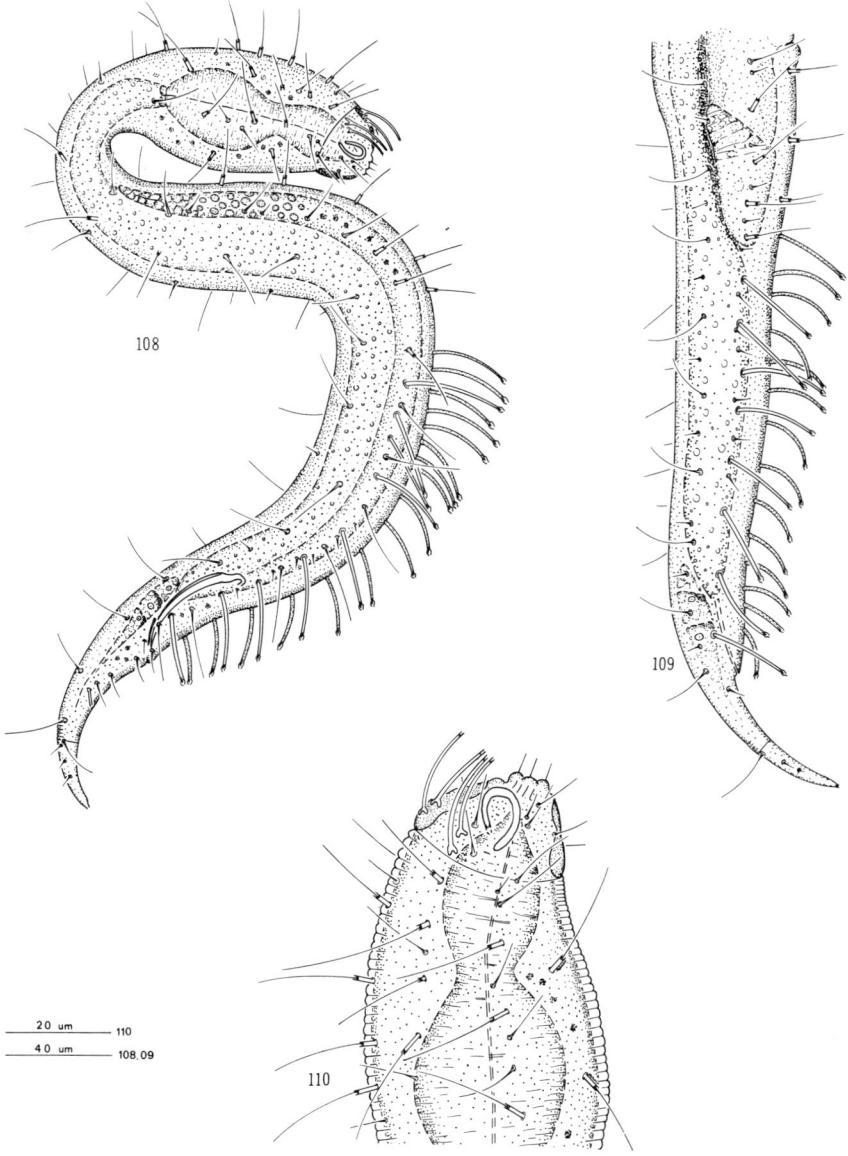

14. Figs. 108-110: *Dracograllus cobbi* n. sp. 108) Male, full length; 109) Female, body region posterior to vulva; 110) Male, head and esophageal region, note pedicel-setae.

Dracograllus mawsoni[13] n. sp.
(Figs. 111 to 112)

Measurements (1 ♀): L = 0.7 mm; b = 6.0; c = 6.0; V = 49%; CAT = 25 μm; No CAT = 8; SER (L/W) = 2.9; SER (E/L) = 21%; SS = 12–45 μm; first SlAT = 37 μm; last SlAT = 28 μm; first SvAT = 34 μm; last SvAT = 27 μm; No SlAT = 15; No SvAT = 16; Non-ann Term to Tail Length = 58%; T/ABD = 6.4.

(2 ♂♂): L = 0.6 (0.5–0.6) mm; a = 13.2 (13.0–13.3); b = 7.3 (6.5–8.0); c = 4.9 (4.7–5.1); CAT = 26 μm; No CAT = 8; SER (L/W) = 2.6 (2.4–2.8); SER (E/L) = 21 (18–23)%; SS = 13–36 μm; first SlAT = 37 (34–41) μm; last SlAT = 26 μm; first SvAT = 31 (27–34) μm; last SvAT = 23 (21–24) μm; No SlAT = 13 (12–14); No SvAT = 13 (12–14); Non-ann Term to Tail Length = 32 (28–35)%; T/ABD = 5.0; Spic = 53 (52–54) μm; Gub = 15 μm.

(Holotype ♂): L = 0.5 mm; a = 13.0; b = 6.5; c = 4.7; CAT = 26 μm; No CAT = 8; SER (L/W) = 2.8; SER (E/L) = 23%; SS = 15–36 μm; first SlAT = 41 μm; last SlAT = 26 μm; first SvAT = 34 μm; last SvAT = 24 μm; No SlAT = 12; No SvAT = 12; Non-ann Term to Tail Length = 35%; T/ABD = 5.0; Spic = 54 μm; Gub = 15 μm.

Male Holotype.—(Figures in parentheses refer to range within species.) Amphids elongate loop-shaped. Eight CAT, 2 sublateral and 2 subdorsal pairs in 2 transverse rows on rostrum, anterior row adjacent to and posterior row posterior to amphids. Annulation without ornamentation. Longest SS on swollen esophageal region. PS[14] in all 8 rows on swollen esophageal region, pedicels 2 to 3 μm long. Two (3) PS on right and 3 (4) on left side of body in dorso-sublateral rows adjacent to anterior end of testis, pedicels 2 to 3 μm long. Two long pairs of PS in dorso-sublateral rows on annulated tail region, 1 about 1 ABD posterior to anus and 1 pair about $1/4$ ABD anterior to last complete tail annule, pedicels 2 μm long. Without stout setae in subventral row just anterior to first SvAT. Long and short setae intermingled with SlAT. Five long setae with SlAT, not alternating with tubes, some setae equal to and others shorter than adjacent tubes. SlAT with 7 (5) long and 7 short tubes, not alternating. All SlAT anterior to anus. Caudal glands extend anterior to anus 2.0 (1.1–2.0) times ABD. Corpus of gubernaculum slightly enlarged laterally. Three (2 to 3) pairs of anal setae; 8 to 14 μm (8 to 16 μm) long, 2 pairs anterior and 1 posterior to anus (when 2 pairs, 1 pair anterior and 1 posterior to anus). Anal flap absent. Two pairs of setae on non-annulated tail region, position measured from last complete tail annule to tail tip; 1 long lateral pair, about 13%, about equal in length to anal setae; 1 short subdorsal (lateral to subdorsal) pair adjacent to long lateral pair (some specimens lateral and 3 to 4 annule widths anterior to long lateral pair); single short subdorsal seta on left side of body about 50%.

Female.—Similar to males. Amphids slightly smaller, with more open loop than in males. Two pairs of subventral paravulval setae, 4 to 7 μm long, 1 anterior and 1 pair posterior to vulva. Longest SS on mid-body region. PS in all 8 rows on swollen esophageal region, pedicels 3 μm long. Three PS on right and 4 on left side of body in dorso-sublateral rows just anterior to distal end of anterior ovary, pedicels 3 μm long. Two long pairs of PS in dorso-sublateral rows on annulated tail region, 1 about 1 $1/2$ ABD posterior to anus and 1 pair about $1/2$ ABD anterior to last complete tail annule, pedicels 2 μm long. Short setae alternate with SlAT. SlAT with 6 long and 9 short tubes. One of the SlAT posterior to anus. Caudal glands extend anterior to anus 1.9 times ABD. Anal flap absent. Two short pairs of setae on non-annulated tail region, 1 lateral pair about 50%, 1 subdorsal pair about 25%.

Second-, Third-, and Fourth-Stage Larvae.—Not observed.

Holotype (♂).—Collected October 1, 1972 by P. Johnston. Catalogue No. UCNC 1423, University of California, Davis.

Paratypes.—1 ♀, 2 ♂♂ (1 ♂ broken). Same data as holotype. Deposited at University of California, Davis (UCNC 1493 to 1495).

Type Habitat.—Marine, associated with bottom debris.
Type Locality.—Long Nose Point, Port Jackson, New South Wales, Australia.
Distribution.—Same as Type Locality.

---

13. Named in honor of Patricia Mawson.
14. All counts and measurements on right side of body unless indicated.

Diagnosis.—Most closely resembles *D. cobbi* n. sp. differs by greater number of SlAT in males; females by greater number of SlAT and 1 SlAT posterior to anus. Differs from *D. demani* n. sp. and *D. kreisi* n. sp. by greater number of SlAT in males; females by greater number of SlAT and 1 SlAT posterior to anus. Differs from *D. stekhoveni* n. sp., *D. gerlachi* n. sp., and *D. falcatus* n. comb. by 8 CAT on rostrum; from remaining known species of *Dracograllus* by PS.

Dracograllus demani[15] n. sp.

(Figs. 113 to 116)

Measurements (20 ♀♀): L = 0.6 (0.5-0.8) mm; b = 6.3 (5.5-7.9); c = 7.7 (6.8-9.3); V = 47 (44-52)%; CAT = 21 (17-26) μm; No CAT = 8; SER (L/W) = 2.5 (2.2-2.9); SER (E/L) = 21 (19-23)%; SS = 14-51 μm; first SlAT = 38 (28-53) μm; last SlAT = 29 (26-34) μm; first SvAT = 36 (31-48) μm; last SvAT = 22 (17-27) μm; No SlAT = 7 (6-8); No SvAT = 12 (10-13); Non-ann Term to Tail Length = 47 (41-51)%; T/ABD = 4.7 (4.1-5.2).

(20 ♂♂): L = 0.6 (0.5-0.8)mm; a = 14.0 (12.2-17.0); b = 7.1 (6.0-8.8); c = 7.6 (6.5-8.8); CAT = 22 (17-27) μm; No CAT = 8; SER (L/W) = 2.6 (2.3-3.0); SER (E/L) = 20 (18-22)%; SS = 14-48 μm; first SlAT = 38 (33-48) μm; last SlAT = 31 (26-39) μm; first SvAT = 35 (30-52) μm; last SvAT = 19 (16-27) μm; No SlAT = 6 (5-7); No SvAT = 10 (8-12); Non-ann Term to Tail Length = 32 (24-39)%; T/ABD = 3.9 (3.5-4.6); Spic = 48 (45-53) μm; Gub = 14 (12-19) μm.

(Holotype ♂): L = 0.6 mm; a = 14.8; b = 7.8; c = 7.8; CAT = 25 μm; No CAT = 8; SER (L/W) = 2.6; SER (E/L) = 19%; SS = 17-41 μm; first SlAT = 46 μm; last SlAT = 32 μm; first SvAT = 38 μm; last SvAT = 19 μm; No SlAT = 6; No SvAT = 10; Non-ann Term to Tail Length = 32%; T/ABD = 3.9; Spic = 49 μm; Gub = 14 μm.

Male Holotype.—(Figures in parentheses refer to range within species.) Amphids elongate loop-shaped. Eight CAT, 2 sublateral and 2 subdorsal pairs in 2 transverse rows on rostrum, anterior row adjacent to and posterior row posterior to amphids. Annulation without ornamentation. Longest SS on swollen esophageal region. PS[16] in all 8 rows on swollen esophageal region, pedicels 2 to 4 μm long. Three PS (3 to 7) on right and 6 (6 to 8) on left side of body in dorso-sublateral rows just anterior to distal end of testis, pedicels 2 to 4 μm long. Two PS (2 to 3) in ventro-sublateral rows just anterior to SlAT, pedicels 2 to 4 μm long. Two PS (1 to 2) on right about mid-body and 1 (1 to 3) on left side of body just anterior to first SvAT in subventral rows, pedicels 2 to 3 μm long. One long pair of PS in dorso-sublateral rows on tail just anterior to last tail annule, pedicels 2 to 3 μm long. Without stout setae in subventral row just anterior to first SvAT. One pair of short, open-ended setae in ventro-sublateral row on annulated tail region, just anterior to last tail annule (fig. 114). Long and short setae intermingled with SlAT. Four (3 to 5) long setae with SlAT, not alternating with tubes, setae either equal to or longer than adjacent tubes (sometimes with 2 long setae occurring together). Anterior 3 long setae with pedicels (usually absent) in SlAT rows. SlAT with 2 (1 to 5) long and 4 (1 to 4) short tubes. All SlAT anterior to anus. Caudal glands extend anterior to anus 1.9 (1.0 to 2.5) times ABD. Corpus of gubernaculum slightly enlarged laterally. Two pairs of anal setae, 6 to 14 μm (6 to 15 μm) long, 1 pair anterior and 1 posterior to anus. Anal flap absent. Two (1 to 4) pairs of setae on non-annulated tail region, position measured from last complete tail annule to tail tip; 1 long lateral (lateral to subventral) pair about 33% equal in length to long setae opposite and dorsal to PAT, 1 (present or absent) short subdorsal pair adjacent to long pair (some specimens with 1 short lateral to subventral pair just posterior to last complete tail annule; if long pair subventral, sometimes 1 short subventral pair present and adjacent to long pair).

Females.—Similar to males. Two paravulval setae, 7 to 9 μm long; either on right or left side of body, never both; usually 1 anterior and 1 posterior to vulva, sometimes setae close together and adjacent to vulva. Longest SS on swollen esophageal region. PS in all 8 rows on swollen esophageal region, pedicels 2 to 4 μm long. Three to 6 PS on right and 5 to 6 on left side of body in dorso-

---

15. Named in honor of J. G. De Man.
16. All counts and measurements on right side of body unless indicated.

sublateral rows about $1/2$ between vulva and distal end of anterior ovary, pedicels 1 to 4 μm long. Three PS in ventro-sublateral rows, 2 just anterior to first SlAT and 1 about $1/2$ between vulva and SlAT, pedicels 2 to 4 μm long. One to 5 PS on right side of body in subventral rows, pedicels 2 μm long; 2 PS present or absent, between vulva and first SvAT; and 1 to 3 PS anterior to vulva. One to 7 PS on left side of body in subventral rows, pedicels 2 μm long; 5 PS present or absent, between vulva and first SvAT; and 1 to 2 PS anterior to vulva. One long pair of PS in dorso-sublateral rows on annulated tail region just anterior to last tail annule, pedicels 1 to 2 μm long. Short setae intermingled with SlAT. Four to 5 long and 2 to 3 short tubes, not alternating. All SlAT anterior to anus. Caudal glands extend anterior to anus 2.1 to 2.6 times ABD. With or without short anal flap. Three pairs of setae on non-annulated tail region; 1 subdorsal pair 1 to 3 annule widths posterior to last complete tail annule or on last annule; 1 subdorsal pair, present or absent, about 33%; 1 lateral to subdorsal pair 50% to 66%.

Second-Stage Larvae.—L = 0.2 to 0.3 mm. Similar to adults. Amphids circular unispiral. CAT typical of genus. PS in all 6 rows on swollen esophageal region, pedicels 1 to 3 μm long. One PS in dorso-sublateral rows just posterior to swollen esophageal region, pedicels 2 to 3 μm long. Two PS in ventro-sublateral rows just anterior to first SlAT, pedicels 2 to 4 μm long. One long pair of PS in dorso-sublateral rows on tail just anterior to last complete tail annule, pedicels 2 to 3 μm long. SlAT typical of genus. Setae on non-annulated tail region, either single dorsal seta or 1 subdorsal pair 33% to 50%.

Third-Stage Larvae.—L = 0.3 to 0.4 mm. Similar to adults. Amphids circular unispiral, ventral arm curving anteriorly against dorsal arm, $1/4$ spiral doubled. CAT typical of genus. PS in all 8 rows on swollen esophageal region, pedicels 1 to 4 μm long. One to 2 PS on right and 1 on left side of body in dorso-sublateral rows just posterior to swollen esophageal region, pedicels 2 to 3 μm long. One to 2 PS in ventro-sublateral rows just anterior to first SlAT, pedicels 2 to 3 μm long. One long pair of PS present or absent, in dorso-sublateral rows on annulated tail region just anterior to last complete tail annule, pedicels 3 to 4 μm long. Five SlAT, usually paired, all tubes anterior to anus. One subdorsal pair of setae on non-annulated tail region 33% to 50%.

Fourth-Stage Larvae.—L = 0.4 to 0.5 mm. Similar to adults. Amphids elongate unispiral. Four CAT on rostrum, 1 sublateral and 1 subdorsal pair. PS in all 8 rows on swollen esophageal region, pedicels 2 to 4 μm long. Three to 4 PS on right and 3 on left side of body in dorso-sublateral rows just posterior to swollen esophageal region, pedicels 1 to 3 μm long. Two PS in ventro-sublateral rows just anterior to first SlAT, pedicels 3 to 5 μm long. Four to 5 PS in ventral row just anterior to first VAT, pedicels 1 to 3 μm long. One long pair of PS in dorso-sublateral rows on annulated tail region just anterior to last complete tail annule, pedicels 2 to 3 μm long. Five pairs of SlAT, all tubes anterior to anus. Usually 6 VAT, some specimens with 7. Two subdorsal pairs of setae on non-annulated tail region, 1 pair 2 to 3 annule widths posterior to last complete tail annule or about 13%, and 1 about 50%.

Holotype (♂).—Collected February 26, 1972 by P. Vitiello. Catalogue No. UCNC 1425, University of California, Davis.

Paratypes.—9 ♀♀, 8 ♂♂. Same data as holotype. Deposited at University of California, Davis (UCNC 1496 to 1497); USDA Nematode Collection, Beltsville, Maryland; U. S. National Museum of Natural History, Smithsonian Institution, Washington, D. C.; Station Marine D'Endoune et Centre D'Océanographie, Marseille, France; and Laboratoria voor Morfologie en Systematiek, Museum voor Dierkunde, Gent, Belgium.

Larval Stages.—2 second, 3 third, 9 fourth. Same data as holotype. Deposited in same nematode collections as paratypes.

Type Habitat.—Marine.

Type Locality.—Near the Chateau d'If, Marseille, France.

Distribution.—Same as type locality.

Diagnosis.—Most closely resembles *D. kreisi* n. sp. differs by PS in ventro-sublateral rows just anterior to SlAT, longer spicules in males, and fewer SlAT in females. Differs from *D. mawsoni* n. sp. by fewer SlAT in males; females by fewer SlAT and SvAT, and all SlAT anterior to anus. Differs from *D. cobbi* n. sp. by fewer SlAT in males; females by fewer SvAT, longer caudal glands, and V at 44% to 52%. Differs from *D. stekhoveni* n. sp., *D. gerlachi* n. sp., and *D. falcatus* n. comb. by 8 CAT on rostrum, and from remaining known species of *Dracograllus* by PS.

Second-stage larvae differ by PS with pedicels 1 to 4 μm long, circular unispiral amphids, number and location of setae on non-annulated tail region. Third-stage larvae differ from other known third stages of *Dracograllus* by PS, and 5 pairs of SlAT. Fourth-stage larvae differ from other known fourth stages of *Dracograllus* by PS, and 5 pairs of SlAT.

Dracograllus filipjevi[17] n. sp.
(Figs. 117 to 120)

Measurements (6 ♀♀): L = 0.6 (0.6–0.7) mm; b = 7.4 (6.0–8.1); c = 8.4 (7.3–10.0); V = 45 (43–48)%; CAT = 17 (14–19) μm; No CAT = 8; SER (L/W) = 3.2 (2.8–3.5); SER (E/L) = 20 (18–21)%; SS = 10–27 μm; first SlAT = 26 (24–28) μm; last SlAT = 22 (18–24) μm; first SvAT = 23 (23–24) μm; last SvAT = 19 (13–21) μm; No SlAT = 13 (12–14); No SvAT = 10 (9–11); Non-ann Term to Tail Length = 53 (46–55)%; T/ABD = 4.6 (4.0–5.2).

(5 ♂♂): L = 0.6 (0.5–0.7) mm; a = 13.5 (11.3–15.0); b = 7.9 (6.5–9.0); c = 9.0 (8.4–10.3); CAT = 17 (14–20) μm; No CAT = 8; SER (L/W) = 3.2 (2.8–3.6); SER (E/L) = 21 (19–23)%; SS = 11–26 μm; first SlAT = 27 (25–30) μm; last SlAT = 22 (20–25) μm; first SvAT = 21 (21–22) μm; last SvAT = 18 (17–20) μm; No SlAT = 9 (8–11); No SvAT = 10 (9–11); Non-ann Term to Tail Length = 43 (40–50)%; T/ABD = 3.6 (2.9–4.0); Spic = 39 (37–40) μm; Gub = 12 (12–13) μm.

(Holotype ♂): L = 0.6 mm; a = 14.3; b = 7.9; c = 9.0; CAT = 18 μm; No CAT = 8; SER (L/W) = 3.0; SER (E/L) = 19%; SS = 11–26 μm; first SlAT = 25 μm; last SlAT = 22 μm; first SvAT = 21 μm; last SvAT = 17 μm; No SlAT = 8; No SvAT = 11; Non-ann Term to Tail Length = 42%; T/ABD = 3.6; Spic = 40 μm; Gub = 12 μm.

Male Holotype.—(Figures in parentheses refer to range within species.) Amphids elongate loop-shaped. Eight CAT, 2 sublateral and 2 subdorsal pairs in 2 transverse rows on rostrum, adjacent to amphids. Annules longitudinally areolated on anterior part of body. Annules with dot-like punctations from swollen mid-body posterior to anus, tail annules without ornamentation. Annules with a few scattered, anteriorly or posteriorly directed minute spines, most prominent on swollen esophageal region. Longest SS on swollen esophageal region. Some setae with cuticular collars, (fig. 118) setae usually on same general body region as PS. Without stout setae in subventral row on body just anterior to first SvAT. Long and short setae intermingled with SlAT. Two (2 to 3) long setae with SlAT, not alternating with tubes, setae shorter (other specimens setae either equal to or shorter) than adjacent tubes. SlAT with 4 (4 to 5) long and 4 (4 to 6) short tubes, not alternating (on some specimens long and short tubes alternate). All SlAT anterior to anus. Caudal glands extend anterior to anus 2.0 (1.4 to 2.0) times ABD. Corpus of gubernaculum laterally enlarged (fig. 120). Two pairs of broad-based anal setae, 8 to 9 μm long, both pairs anterior (some specimens with 1 anterior and 1 pair adjacent) to anus. Anal flap absent. Three pairs of setae on non-annulated tail region, position measured from last complete tail annule to tail tip; 1 subventral pair about 25% (25% to 50%); 2 lateral (lateral to subdorsal) pairs, one just posterior to last complete tail annule (posterior to or on last tail annule), and 1 slightly posterior to subventral pair (some specimens adjacent to subventral pair or about 66%).

Females.—Similar to males. Two pairs of paravulval setae, 5 to 6 μm long, 1 pair anterior and 1 posterior to vulva. Longest SS on swollen esophageal region. Short setae intermingled with SlAT. SlAT with 3 to 5 long and 8 to 9 short tubes. One of the SlAT posterior to anus. Caudal glands extend anterior to anus 1.0 to 3.6 times ABD. Two to 5 pairs of setae on non-annulated tail region; 1 long lateral to subdorsal pair, usually subdorsal, just posterior to or on last complete tail annule, about equal in length to long setae on annulated tail region; 1 short lateral pair 50% to 66%; 1 short lateral to subdorsal pair, present or absent, about 33%; 2 short lateral pairs present or absent, 50% and 66%.

Second- and Third-Stage Larvae.—Not observed.

Fourth-Stage Larvae.—L = 0.5 to 0.7 mm. Similar to adults. Amphids elongate loop-shaped, amphidial arms converged almost forming elongated unispiral. Four CAT, 1 sublateral and 1 subdorsal pair in 2 transverse rows on rostrum. Seven to 8 pairs of SlAT, with 1 of the SlAT posterior

17. Named in honor of I. N. Filipjev.

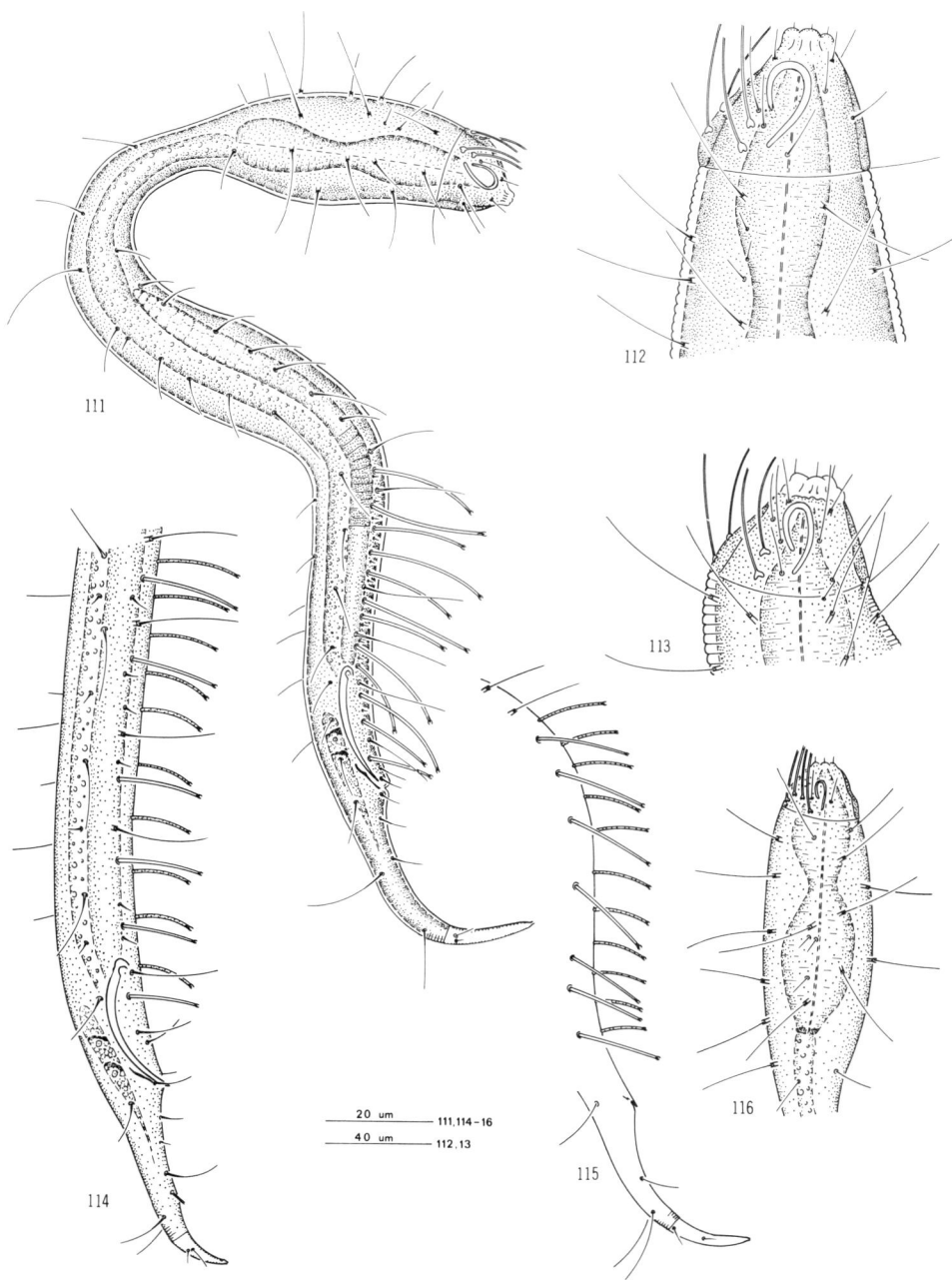

15. Figs. 111-112: *Dracograllus mawsoni* n. sp. 111) Male, full length, SvAT not illustrated; 112) Male, head. Figs. 113-116: *Dracograllus demani* n. sp. 113) Male, head; 114) Male, posterior body region; 115) Female, PAT[a] and tail; 116) Male, head and esophageal region.
a. Short SS with SlAT not illustrated.

to anus. Eight to 9 VAT. Three pairs of setae on non-annulated tail region; 1 long subdorsal pair just posterior to last complete tail annule about equal in length to long setae on annulated tail region; 1 short subdorsal pair about 25%; 1 short lateral to subdorsal, usually lateral, pair about 50%.

Holotype (♂).—Collected June 25, 1967 by I. M. Newell and T. Imamura. Catalogue No. UCNC 1424, University of California, Davis.

Paratypes.—7 ♀♀ (1 ♀ broken), 5 ♂♂. Same data as holotype. Deposited at University of California, Davis (UCNC 1501 to 1502); USDA Nematode Collection, Beltsville, Maryland; and Laboratoria voor Morfologie en Systematiek, Museum voor Dierkunde, Gent, Belgium.

Larval Stages.—3 fourth. Same data as holotype. Deposited at University of California, Davis.
Type Habitat.—Marine, associated with holdfasts of Kelp.
Type Locality.—Oarai, Ibaraki-ken, Honshu Island, Japan.
Distribution.—Same as type locality.

Diagnosis.—Most closely resembles *D. timmi* n. sp. differs by annules ornamented with few, scattered, minute spines. Differs from *D. stekhoveni* n. sp., *D. gerlachi* n. sp., and *D. falcatus* n. comb. by 8 CAT on rostrum. Differs from *D. cobbi* n. sp., *D. mawsoni* n. sp., *D. demani* n. sp., and *D. kreisi* n. sp. by absence of PS; and from *D. eira* n. comb. by CAT adjacent to amphids. Differs from *D. chitwoodi* n. sp., *D. wieseri* n. sp., and *D. solidus* n. comb. by setae with cup-shaped cuticular collars.

Fourth-stage most closely resemble fourth-stage *D. timmi* n. sp. differ by 3 pairs of setae on non-annulated tail region, absence of stout setae in ventral row just anterior to first VAT, and fewer VAT. Differ from *D. gerlachi* n. sp. and *D. stekhoveni* n. sp. by 4 CAT on rostrum, fewer SlAT and VAT; and from *D. demani* n. sp. by absence of PS, greater number of SlAT, and 1 SlAT posterior to anus.

Dracograllus timmi[18] n. sp.

(Figs. 121 to 125)

Measurements (16 ♀♀): L = 0.5 (0.5–0.6) mm; b = 5.7 (5.0–6.8); c = 7.8 (5.6–10.3); V = 48 (45–50)%; CAT = 18 (16–20) μm; No CAT = 8; SER (L/W) = 3.1 (2.6–3.3); SER (E/L) = 23 (21–26)%; SS = 11–37 μm; first SlAT = 25 (21–29) μm; last SlAT = 21 (17–25) μm; first SvAT = 21 (18–24) μm; last SvAT = 17 (15–21) μm; No SlAT = 10 (9–12); No SvAT = 9 (7–11); Non-ann Term to Tail Length = 48 (43–52)%; T/ABD = 3.9 (3.7–4.2).

(20 ♂♂): L = 0.6 (0.5–0.7) mm; a = 11.7 (10.3–13.3); b = 6.0 (5.5–7.1); c = 8.2 (7.0–10.3); CAT = 19 (17–21) μm; No CAT = 8; SER (L/W) = 3.2 (2.7–3.7); SER (E/L) = 22 (20–25)%; SS = 12–38 μm; first SlAT = 26 (24–28) μm; last SlAT = 23 (18–27) μm; first SvAT = 22 (19–23) μm; last SvAT = 16 (13–20) μm; No SlAT = 8 (7–10); No SvAT = 9 (7–13); Non-ann Term to Tail Length = 32 (29–36)%; T/ABD = 3.3 (2.8–3.8); Spic = 46 (41–51) μm; Gub = 15 (13–17) μm.

(*Holotype* ♂): L = 0.6 mm; a = 11.0; b = 5.7; c = 8.1; CAT = 19 μm; No CAT = 8; SER (L/W) = 3.1; SER (E/L) = 23%; SS = 13–35 μm; first SlAT = 25 μm; last SlAT = 23 μm; first SvAT = 23 μm; last SvAT = 18 μm; No SlAT = 8; No SvAT = 11; Non-ann Term to Tail Length = 31%; T/ABD = 3.3; Spic = 47 μm; Gub = 17 μm.

Male Holotype.—(Figures in parentheses refer to range within species.) Amphids elongate loop-shaped (some specimens ventral arm curved anteriorly toward dorsal arm almost forming unispiral). Eight CAT, 2 sublateral and 2 subdorsal pairs in 2 transverse rows on rostrum, adjacent to amphids. Annules with faint annular ridges and spine-like projections, appearing as 2 rows of fine punctations, similar to fig. 166. Tail annules without ornamentation. Longest SS on mid-body region (some specimens longest both on swollen esophageal and mid-body regions). Some setae with cuticular collars (fig. 122) occurring in same general body region as PS. Stout setae in subventral rows just anterior to first SvAT, 3 (3 to 8) on right and 2 (2 to 4) on left side of body, distal ends directed posteriorly. One pair of short, stubby, open-ended setae in ventro-sublateral row on tail just anterior

---

18. Named in honor of R. W. Timm.

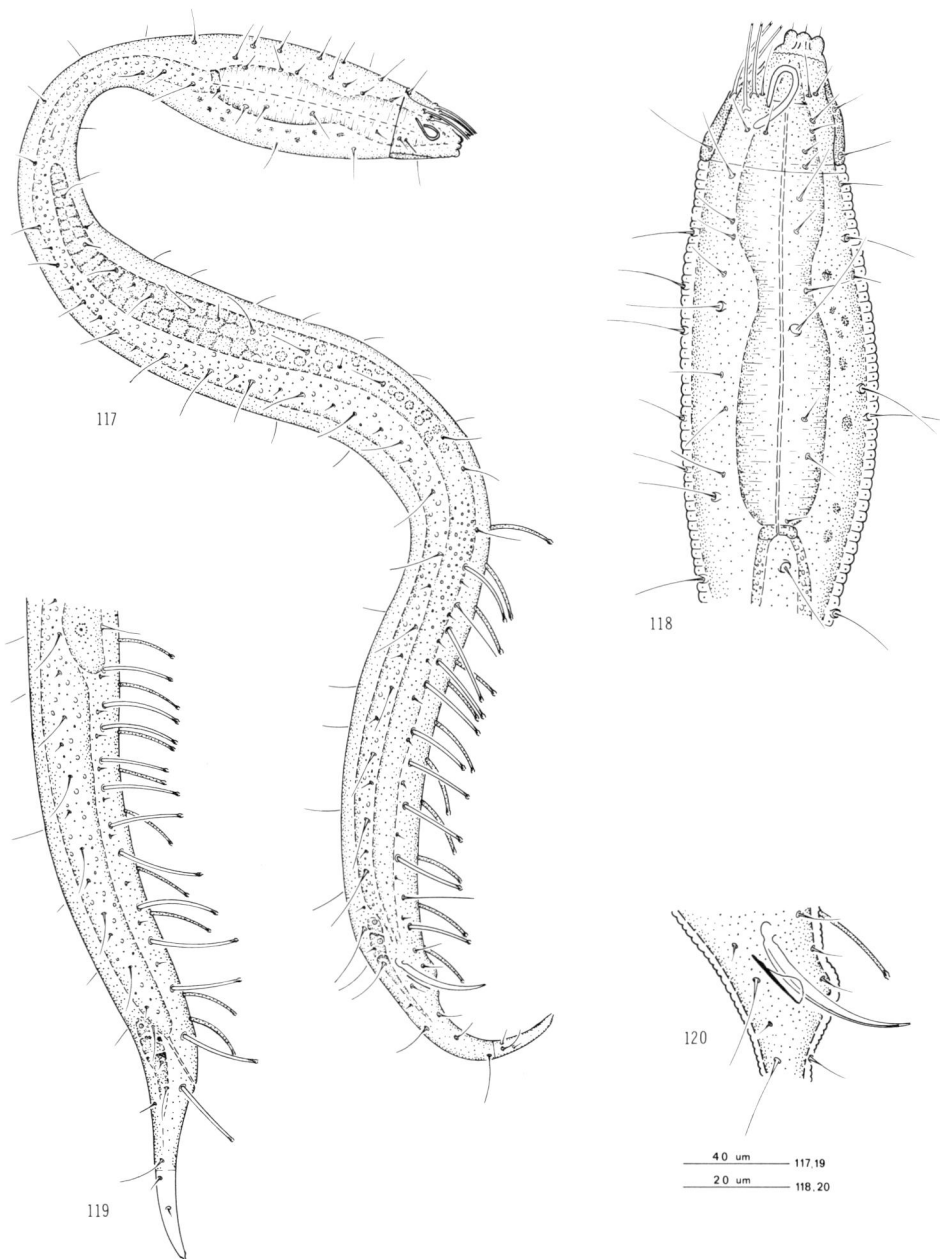

16. Figs. 117-120: *Dracograllus filipjevi* n. sp. 117) Male, full length; 118) Male, head and esophageal region; 119) Female, posterior body region; 120) Male, anal region.

to last tail annule (fig. 123). Long and short setae intermingled with SlAT. Four (3 to 6) long setae with SlAT, not alternating with tubes, setae either equal to or shorter than adjacent tubes. SlAT with 2 (1 to 4) long and 6 (3 to 8) short tubes. All SlAT anterior to anus. Caudal glands extend anterior to anus 2.1 (2.1 to 3.2) times ABD. Corpus of gubernaculum laterally enlarged (fig. 123). Five pairs of anal setae, 8 to 10 µm (7 to 10 µm) long; 2 pairs anterior, 1 adjacent to, and 2 posterior to anus. Anal flap absent. One pair of open-ended setae in sublatero-ventral row just anterior to non-annulated tail region. Two (1 to 2) pairs of setae on non-annulated tail region, position measured from last complete tail annule to tail tip; 1 subventral pair (lateral to subventral, usually subventral) 3 annule widths posterior to last complete tail annule (some specimens just posterior to last complete tail annule); 1 latero-subventral pair (present or absent, either lateral, subdorsal, or latero-subventral) about 50%.

Females. —Similar to males. Amphids elongate loop-shaped. Two pairs of paravulval setae, 5 to 7 µm long, 1 anterior and 1 pair posterior to vulva. Longest SS on mid-body region. Stout setae in subventral rows just anterior to first SvAT, 2 to 4 on right and 1 to 3 on left side of body. Short setae intermingled with SlAT. SlAT with 2 to 5 long and 6 to 9 short tubes. One of the SlAT posterior to anus. Caudal glands extend anterior to anus 1.8 to 2.8 times ABD. Anal flap short. Three subdorsal pairs of setae on non-annulated tail region, 1 just posterior to or on last complete tail annule, 1 about 25%, and 1 about 66%.

Second- and Third-Stage Larvae. —Not observed.

Fourth-Stage Larvae. —L = 0.4 to 0.5 mm. Similar to adults. Amphids elongate unispiral. Four CAT, 1 sublateral and 1 subdorsal pair in 2 transverse rows on rostrum. Four stout setae in ventral row just anterior to first VAT. Seven pairs of SlAT, 1 of the SlAT posterior to anus. Seven VAT. Two pairs of setae on non-annulated tail region; 1 subdorsal pair just posterior to last complete tail annule, 1 lateral pair about 50%.

Holotype (♂).—Collected January 29, 1968 by I. M. Newell. Catalogue No. UCNC 1426, University of California, Davis.

Paratypes.—18 ♀♀, 33 ♂♂. Same data as holotype. Deposited at University of California, Davis (UCNC 1504 to 1506); USDA Nematode Collection, Beltsville, Maryland; U. S. National Museum of Natural History, Smithsonian Institution, Washington, D. C.; Station Marine D'Endoune et Centre D'Océanographie, Marseille, France; and Laboratoria voor Morfologie en Systematiek, Museum voor Dierkunde, Gent, Belgium.

Larval Stages.—2 fourth. Same data as holotype. Deposited at University of California, Davis.

Type Habitat. —Marine, from coarse sand.

Type Locality. —East shore near Motu Toopau, Bora Bora Island, Society Islands.

Distribution. —Same as type locality.

Diagnosis. —Most closely resembles *D. filipjevi* n. sp. differs by faint annular ridges with spine-like projections appearing as 2 rows of fine punctations. Differs from *D. stekhoveni* n. sp., *D. gerlachi* n. sp., and *D. falcatus* n. comb. by 8 CAT on rostrum; and from *D. mawsoni* n. sp., *D. cobbi* n. sp., *D. demani* n. sp., and *D. kreisi* n. sp. by absence of PS. Differs from *D. eira* n. comb. by CAT adjacent to amphids; and from *D. solidus* n. comb., *D. chitwoodi* n. sp., and *D. wieseri* n. sp. by some SS with cup-shaped cuticular collars.

Fourth-stage most closely resemble fourth-stage *D. filipjevi* n. sp. differ by 2 pairs of setae on non-annulated tail region, presence of stout setae in ventral row just anterior to first VAT, and fewer VAT. Differ from *D. gerlachi* n. sp. and *D. stekhoveni* n. sp. by 4 CAT on rostrum; and from *D. demani* n. sp. by absence of PS, and greater number of SlAT.

Dracograllus chitwoodi[19] n. sp.

(Figs. 126, 128)

Measurements (3 ♀♀): L = 0.5 (0.5-0.6) mm; b = 5.7 (5.3-6.1); c = 7.1 (6.1-8.0); V = 45 (43-48)%; CAT = 19 (17-20) µm; No CAT = 8; SER (L/W) = 2.8 (2.5-3.0); SER (E/L) = 23 (22-24)%; SS = 12-33 µm; first SlAT = 21 (20-22) µm; last SlAT = 28 (26-31) µm; first SvAT = 19 (18-20)

---

19. Named in honor of B. G. Chitwood.

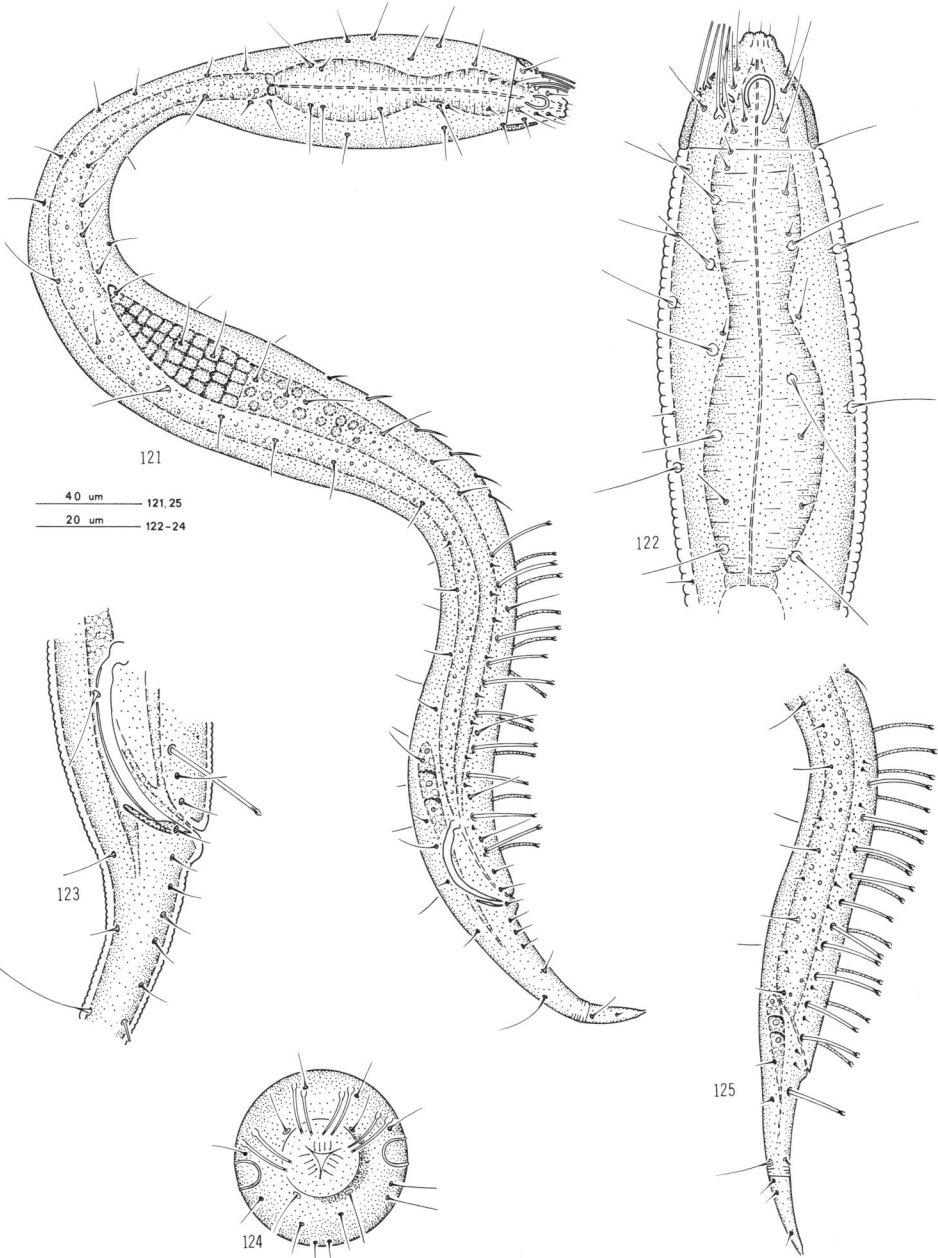

17. Figs. 121–125: *Dracograllus timmi* n. sp. 121) Male, full length; 122) Male, head and esophageal region; 123) Male, anal region; 124) Male, face view; 125) Female, posterior body region.

μm; last SvAT = 23 (22-25) μm; No SlAT = 10 (9-10); No SvAT = 8 (8-10); Non-ann Term to Tail Length = 52 (50-54)%; T/ABD = 5.6 (5.3-5.8).

(Holotype ♀): L = 0.5 mm; b = 5.7; c = 7.3; V = 43%; CAT = 20 μm; No CAT = 8; SER (L/W) = 3.0; SER (E/L) = 24%; SS = 12-33 μm; first SlAT = 20 μm; last SlAT = 26 μm; first SvAT = 19 μm; last SvAT = 25 μm; No SlAT = 9; No SvAT = 8; Non-ann Term to Tail Length = 54%; T/ABD = 5.6.

Female Holotype.—(Figures in parentheses refer to range within species.) Amphids elongate loop-shaped. Eight CAT, 2 sublateral and 2 subdorsal pairs in 2 transverse rows on rostrum, anterior row adjacent to and posterior row posterior to amphids. Annules ornamented with 2 transverse rows of dot-like punctations to anus. Annules with a few scattered, posteriorly directed minute spines; tail annules without ornamentation. Paravulval setae absent. Longest SS on mid-body region (some specimens longest on both swollen esophageal and mid-body regions). Some setae with indistinct cuticular collars occurring in same general body region as PS. Without stout setae in subventral rows just anterior to first SvAT. (Some specimens with 1 pair of short, stubby, open-ended setae in ventro-sublateral rows on tail anterior to last annule.) Short setae intermingled with SlAT. SlAT with 4 long and 5 short tubes (4 long and 6 short tubes), long and short tubes alternating (usually alternating). One of the SlAT posterior to anus. Caudal glands extend anterior to anus 3.1 (1.9 to 3.1) times ABD. Anal flap short. Three pairs of setae on non-annulated tail region, position measured from last complete tail annule to tail tip; 1 subdorsal pair about 13% (13% to 25%), 1 subdorsal (lateral to subdorsal) about 66% (50% to 66%), and 1 lateral (lateral to subventral) about 25% (some specimens about 25% or adjacent to most anterior subdorsal pair).

Males.—Not observed.

Second-, Third-, and Fourth-Stage Larvae.—Not observed.

Holotype (♀).—Collected February 20, 1965 by B. E. Hopper. Catalogue No. UCNC 1427, University of California, Davis.

Paratypes.—2 ♀♀. Same data as holotype. Deposited at University of California, Davis, California (UCNC 1507 to 1508).

Type Habitat.—Marine, associated with *Halimeda* sp., a calcareous alga.

Type Locality.—Coral Key, Florida, USA.

Distribution.—Same as type locality.

Diagnosis.—Most closely resembles females of *D. wieseri* n. sp. differs by fewer SlAT and SvAT. Differs from *D. stekhoveni* n. sp., *D. gerlachi* n. sp., and *D. falcatus* n. comb. by 8 CAT on rostrum; and from *D. mawsoni* n. sp., *D. cobbi* n. sp., *D. demani* n. sp., and *D. kreisi* n. sp. by absence of PS. Differs from *D. eira* n. comb. by CAT adjacent to and posterior to amphids; and from *D. solidus* n. comb. by elongate loop-shaped amphids and 1 SlAT posterior to anus. Differs from *D. timmi* n. sp. and *D. filipjevi* n. sp. by absence of setae with cup-shaped cuticular collars.

Dracograllus kreisi[20] n. sp.

(Figs. 127, 129 to 130)

Measurements (1 ♀): L = 0.4 mm; b = 6.0; c = 5.1; V = 50%; CAT = 12 μm; No CAT = 8; SER (L/W) = 2.2; SER (E/L) = 22%; SS = 13-19 μm; first SlAT = 10 μm; last SlAT = 16 μm; first SvAT = 10 μm; last SvAT = 11 μm; No SlAT = 12; No SvAT = 9; Non-ann Term to Tail Length = 69%; T/ABD = 5.3.

(Holotype ♂): L = 0.4 mm; a = 11.8; b = 6.7; c = 6.7; CAT = 14 μm; No CAT = 8; SER (L/W) = 1.9; SER (E/L) = 20%; SS = 11-20 μm; first SlAT = 28 μm; last SlAT = 22 μm; first SvAT = 29 μm; last SvAT = 14 μm; No SlAT = 5; No SvAT = 11; Non-ann Term to Tail Length = 40%; T/ABD = 4.0; Spic = 36 μm; Gub = 10 μm.

Male Holotype.—Amphids elongate loop-shaped. Eight CAT, 2 sublateral and 2 subdorsal pairs in 2 transverse rows on rostrum, anterior row adjacent to and posterior row just anterior to first body annule. Annules faintly ornamented with 2 transverse rows of dot-like punctations on swollen

---

20. Named in honor of H. A. Kreis.

mid-body region. Longest SS on swollen esophageal region. PS[21] in all 8 rows on swollen esophageal region, pedicels 1 to 2 μm long. Two PS on right and 1 on left side of body in dorso-sublateral rows adjacent to anterior end of testis, pedicels 2 μm long. Two PS in ventro-sublateral rows just anterior to first SlAT, pedicels 1 μm long. Two PS in subventral rows just anterior to first SvAT, pedicels 1 to 2 μm long. Without stout setae in subventral rows just anterior to first SvAT. Long and short setae intermingled with SlAT. Four long setae with SlAT, alternating with tubes, setae either equal to or shorter than adjacent tubes. SlAT with 4 long and 1 short tube. All SlAT anterior to anus. Caudal glands extend anterior to anus 1.4 times ABD. Two pairs of anal setae, 5 to 6 μm long, 1 anterior and 1 pair posterior to anus. Anal flap absent. Four pairs of setae on non-annulated tail region, position measured from last complete tail annule to tail tip; 1 short lateral pair about 25%; 1 long lateral pair adjacent to and slightly ventral to short lateral pair about equal in length to long setae on annulated tail region; 2 subdorsal pairs, 1 long pair just posterior to last complete tail annule, and 1 short pair, seta on right side of body about 13% and seta on left side of body about 50%.

Female.—Similar to male. Annulation without ornamentation. Paravulval setae absent. Longest SS on rostrum. PS in all 8 rows on swollen esophageal region, pedicels 1 to 2 μm long. Two PS in dorso-sublateral rows on swollen mid-body region and 2 PS in subventral rows just anterior to first SvAT, pedicels 1 μm long. Short setae intermingled with SlAT. SlAT with 6 long and 6 short tubes, not alternating. All SlAT anterior to anus. Caudal glands extend anterior to anus 2.9 times ABD. Four pairs of setae on non-annulated tail region; 1 lateral pair about 25%; 3 lateral to subdorsal pairs, 1 just posterior to last complete tail annule, 1 about 13%, and 1 about 50%.

Second-, Third-, and Fourth-Stage Larvae.—Not observed.

Holotype (♂).—Collected June 3, 1968 by I. M. Newell. Catalogue No. UCNC 1428, University of California, Davis.

Paratype.—1 ♀. Same data as holotype. Deposited at University of California, Davis (UCNC 1509).

Type Habitat.—Marine, associated with *Halimeda* sp., a calcareous alga, growing on coral reef.

Type Locality.—Near Coco Solo, on Galeta Beach, Panama; in the Caribbean Sea.

Distribution.—Same as type locality.

Diagnosis.—Most closely resembles *D. demani* n. sp. differs by absence of PS in ventro-sublateral rows just anterior to first SlAT; shorter spicules in males; and greater number of SlAT in females. Differs from *D. mawsoni* n. sp. and *D. cobbi* n. sp. by fewer SlAT in males, and fewer SvAT in females; from *D. stekhoveni* n. sp., *D. gerlachi* n. sp., and *D. falcatus* n. comb. by 8 CAT on rostrum; and from *D. eira* n. comb. by CAT adjacent to and posterior to amphids. Differs from other known species of *Dracograllus* by PS.

Dracograllus gerlachi[22] n. sp.

(Figs. 131 to 134)

Measurements (1 ♀): L = 0.7 mm; b = 6.8; c = 8.5; V = 50%; CAT = 23 μm; No CAT = 13; SER (L/W) = 2.8; SER (E/L) = 22%; SS = 7-28 μm; first SlAT = 28 μm; last SlAT = 17 μm; first SvAT = 25 μm; last SvAT = 14 μm; No SlAT = 24; No SvAT = 21; Non-ann Term to Tail Length = 40%; T/ABD = 3.3.

(Holotype ♂): L = 0.6 mm; a = 11.3; b = 7.0; c = 9.0; CAT = 22 μm; No CAT = 13; SER (L/W) = 2.6; SER (E/L) = 22%; SS = 7-30 μm; first SlAT = 21 μm; last SlAT = 24 μm; first SvAT = 19 μm; last SvAT = 14 μm; No SlAT = 13; No SvAT = 18; Non-ann Term to Tail Length = 28%; T/ABD = 2.5; Spic = 39 μm; Gub = 18 μm.

Male Holotype.—Amphids elongate loop-shaped. Thirteen CAT; 2 sublateral and 3 subdorsal pairs, 3 unpaired dorsals; in 3 transverse rows on rostrum, anterior row adjacent to and posterior row posterior to amphids. Annules ornamented with dot-like punctations, most prominent on mid-body region; tail annules without ornamentation. Annules with a few scattered, anteriorly or pos-

---

21. All counts and measurements on right side of body unless indicated.
22. Named in honor of S. A. Gerlach.

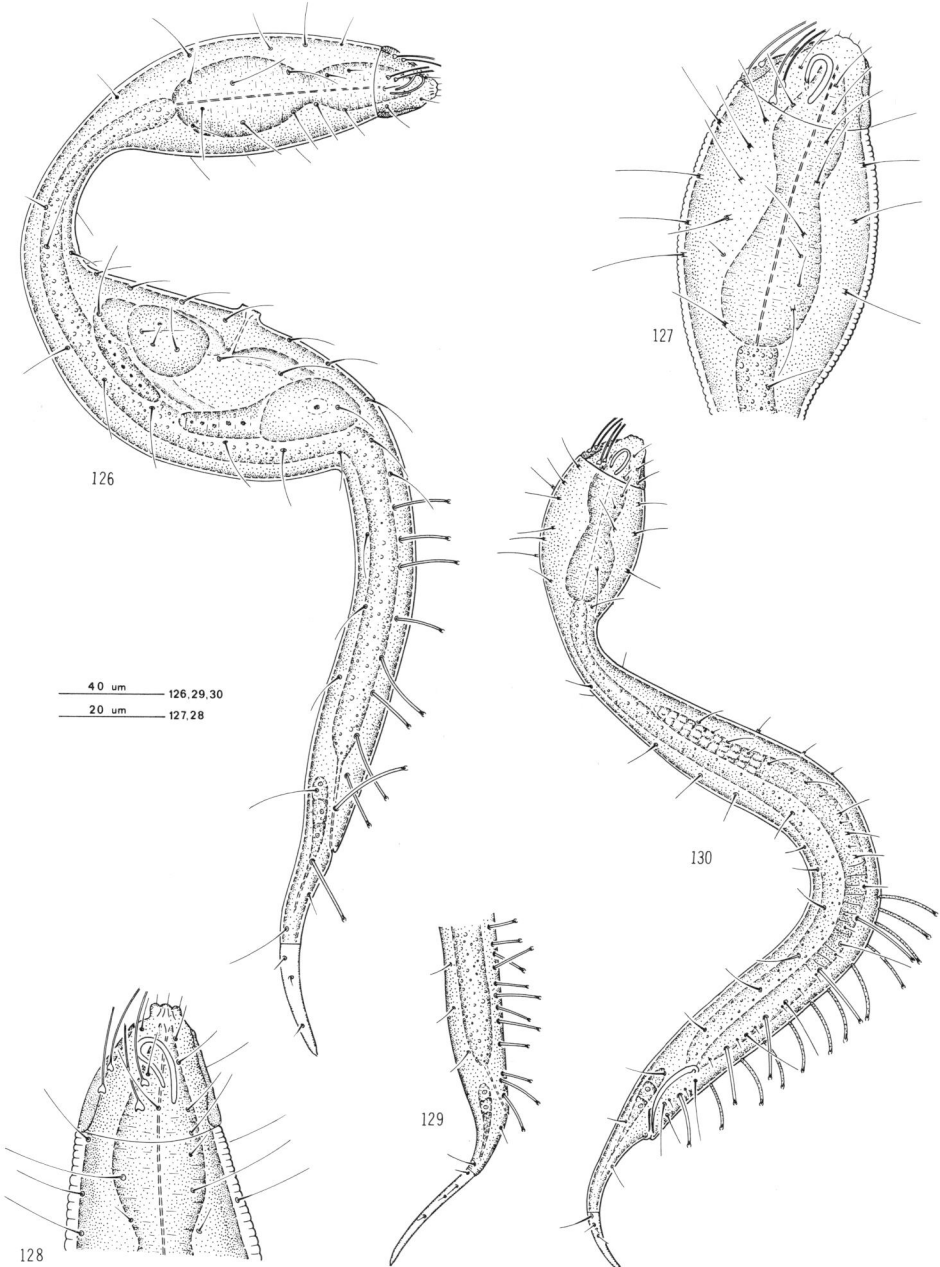

18. Figs. 126, 128: *Dracograllus chitwoodi* n. sp. 126) Female, full length[a], SvAT not illustrated; 128) Female, head. Figs. 127, 129–130: *Dracograllus kreisi* n. sp. 127) Male, head and esophageal region; 129) Female, posterior body region[a], SvAT not illustrated; 130) Male, full length.

a. Short SS with SlAT not illustrated.

teriorly directed minute spines. Longest SS on swollen esophageal region. Stout setae in subventral rows just anterior to first SvAT, 8 on right and 7 on left side of body, distal ends directed posteriorly. Long and short setae intermingled with SlAT. SlAT with 7 long setae, not alternating with tubes, setae variable lengths. SlAT with 8 long and 5 short tubes. All SlAT anterior to anus. Caudal glands extend anterior to anus 1.2 times ABD. Corpus of gubernaculum enlarged laterally (fig. 133). Two pairs of broad-based anal setae, 7 to 8 µm long, 1 pair adjacent to and 1 posterior to anus. Anal flap absent. Two pairs of setae on non-annulated tail region, position measured from last complete tail annule to tail tip; 1 long subventral pair 2 annule widths posterior to last complete tail annule, about equal in length to long setae in sublatero-ventral rows on annulated tail region, and 1 short lateral pair about 50%.

Female.—Similar to male. Annulation ornamentation less prominent than male. Two pairs of paravulval setae, 5 µm long, 1 pair anterior and 1 posterior to vulva. Longest SS on swollen esophageal and mid-body regions. Ventro-sublateral rows of SS on body with only short, stubby setae. Stout setae in subventral rows just anterior to first SvAT, 11 on right and 10 on left side of body. Short setae intermingled with SlAT. SlAT with 11 long and 13 short tubes, not alternating. One of the SlAT adjacent to anus and 2 posterior to anus. Caudal glands extend anterior to anus 3.1 times ABD. Anal flap short. Two pairs of setae on non-annulated tail region, 1 subdorsal pair about 33%, and 1 lateral pair about 66%.

Second- and Third-Stage Larvae.—Not observed.

Fourth-Stage Larvae.—L = 0.6 mm. Similar to adults. Amphids elongate loop-shaped. Six CAT, 1 sublateral and 2 subdorsal pairs in 2 transverse rows on rostrum. Six stout setae in ventral row just anterior to first VAT. Eleven pairs of SlAT, 1 of the SlAT adjacent to and 1 posterior to anus. Eleven VAT. Two pairs of setae on non-annulated tail region, 1 subdorsal pair just posterior to last complete tail annule, and 1 lateral pair about 50%.

Holotype (♂).—Collected August 13, 1967 by I. M. Newell. Catalogue No. UCNC 1429, University of California, Davis.

Paratype.—1 ♀. Same data as holotype. Deposited at University of California, Davis (UCNC 1510).

Larval Stages.—1 fourth. Same data as holotype. Deposited at University of California, Davis.

Type Habitat.—Marine, associated with brown algae growing on rocks.

Type Locality.—Ibusuki, Kyushu Island, Japan.

Distribution.—Same as type locality.

Diagnosis.—Most closely resembles *D. falcatus* n. comb. differs by longer swollen esophageal region, greater number of stout setae in subventral rows just anterior to first SvAT and shorter spicules in males; greater number of SlAT and 2 SlAT posterior to anus in females. Differs from *D. stekhoveni* n. sp. by absence of Ceph Acan-set on rostrum, and from other known species of *Dracograllus* by 13 CAT on rostrum.

Fourth-stage most closely resemble fourth-stage *D. stekhoveni* n. sp. differ by absence of Ceph Acan-set on rostrum. Differ from *D. demani* n. sp., *D. filipjevi* n. sp., and *D. timmi* n. sp. by 6 CAT on rostrum.

Dracograllus stekhoveni[23] n. sp.

(Figs. 135 to 139)

Measurements (14 ♀♀): L = 0.6 (0.5–0.6) mm; b = 6.1 (5.6–6.9); c = 7.3 (6.0–8.7); V = 49 (45–52)%; CAT = 18 (15–20) µm; No CAT = 14–15; SER (L/W) = 2.5 (2.0–2.9); SER (E/L) = 22 (20–25)%; SS = 4–37 µm; first SlAT = 22 (20–25) µm; last SlAT = 17 (16–19) µm; first SvAT = 19 (13–23) µm; last SvAT = 14 (12–16) µm; No SlAT = 32 (24–37); No SvAT = 24 (21–29); Non-ann Term to Tail Length = 42 (37–47)%; T/ABD = 3.9 (3.4–4.2); Ceph Acan-set = 3 (2–4) µm.

(23 ♂♂): L = 0.5 (0.5–0.6) mm; a = 9.5 (8.7–10.6); b = 6.3 (5.5–7.3); c = 8.0 (6.4–10.0); CAT = 18 (15–23) µm; No CAT = 14–15; SER (L/W) = 2.4 (2.2–2.9); SER (E/L) = 21 (20–25)%; SS = 5–34 µm; first SlAT = 19 (17–24) µm; last SlAT = 17 (15–21) µm; first SvAT = 16 (14–18)

23. Named in honor of J. H. Schuurmans Stekhoven, Jr.

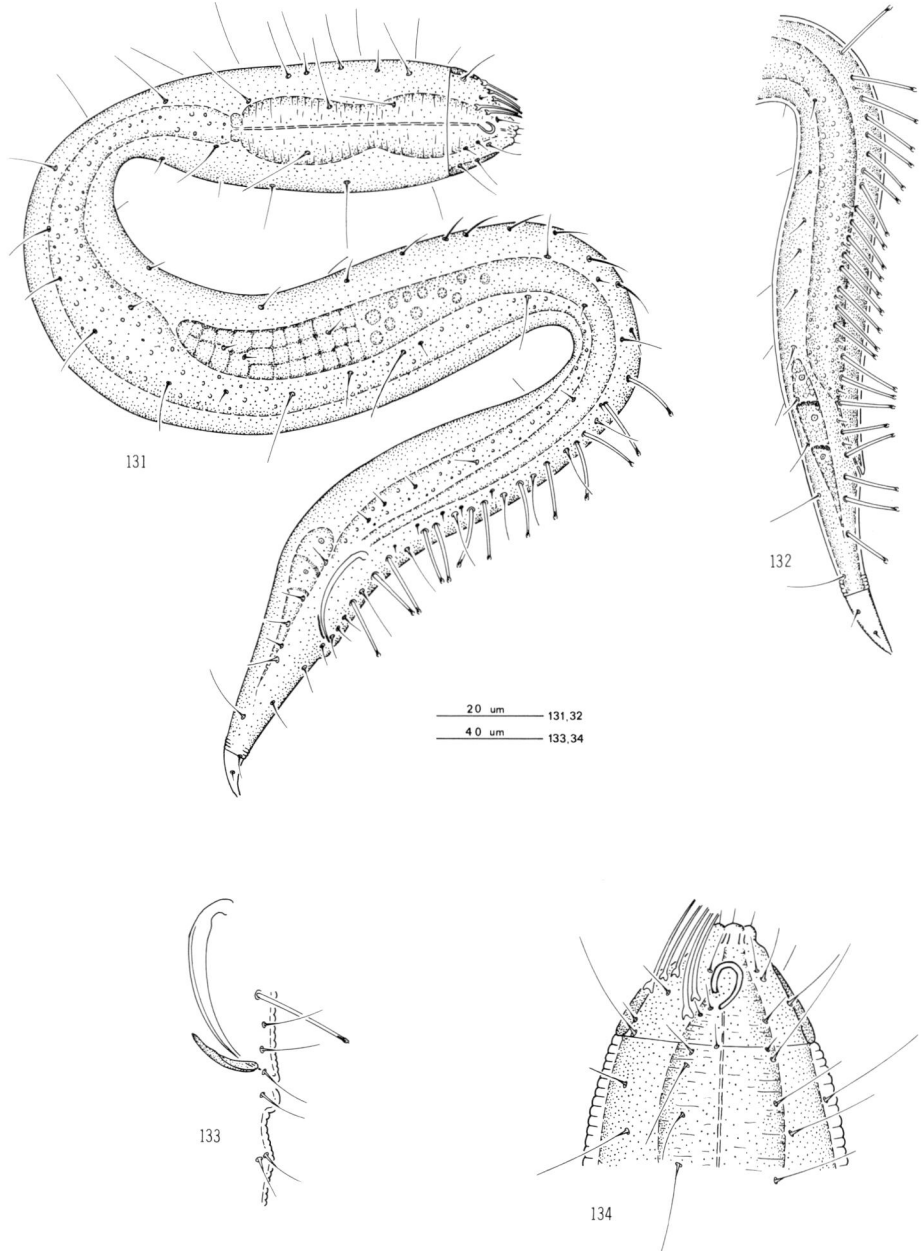

19. Figs. 131–134: *Dracograllus gerlachi* n. sp. 131) Male, full length, SvAT and correct number of CAT not illustrated; 132) Female, posterior body region[a], SvAT not illustrated; 133) Male, anal region; 134) Male, head, correct number of CAT not illustrated.

a. Short SS with SlAT not illustrated.

μm; last SvAT = 12 (10–15) μm; No SlAT = 20 (16–23); No SvAT = 19 (16–26); Non-ann Term to Tail Length = 31 (22–34)%; T/ABD = 2.9 (2.7–3.2); Spic = 44 (40–50) μm; Gub = 17 (15–19) μm; Ceph Acan-set = 3 (2–4) μm.

(Holotype ♂): L = 0.5 mm; a = 9.8; b = 6.1; c = 8.2; CAT = 19 μm; No CAT = 14; SER (L/W) = 2.2; SER (E/L) = 20%; SS = 12–26 μm; first SlAT = 19 μm; last SlAT = 17 μm; first SvAT = 17 μm; last SvAT = 12 μm; No SlAT = 18; No SvAT = 19; Non-ann Term to Tail Length = 34%; T/ABD = 2.9; Spic = 44 μm; Gub = 17 μm; Ceph Acan-set = 3 μm.

Male Holotype. —(Figures in parentheses refer to range within species.) Amphids elongate loop-shaped. Fourteen (15) CAT; 2 sublateral, 3 dorso-sublateral and 2 subdorsal pairs (15 CAT, 2 sublateral, 3 dorso-sublateral and 2 subdorsal pairs, with single dorsal tube); in 3 transverse rows on rostrum, anterior row adjacent to and posterior row posterior to amphids (fig. 137). One pair of sublateral Ceph Acan-set slightly ventral at mid-rostrum (mid-rostrum to anterior $1/3$ of rostrum), 6 (5 to 10) annule widths or 8 μm (7 to 9 μm) anterior to first body annule. Annulation without ornamentation. Longest SS on mid-body region (usually on swollen esophageal region). Stout, posteriorly directed, setae in subventral rows just anterior to first SvAT, 8 (7 to 8) on right and 6 on left side of body. Long and short setae intermingled with SlAT. Five (5 to 9) long setae with SlAT, not alternating with tubes, setae either equal to or longer than adjacent tubes. SlAT with 10 long and 8 short tubes (8 to 16 long and 5 to 12 short tubes). All SlAT anterior to anus. Caudal glands extend anterior to anus 1.6 (1.6 to 3.5) times ABD. Four pairs of broad-based anal setae, 8 to 9 μm (7 to 10 μm) long, 2 pairs anterior and 2 posterior to anus (usually 1 pair anterior, 1 adjacent to, and 2 posterior to anus). Anal flap absent. Three (2 to 3) pairs of setae on non-annulated tail region, position measured from last complete tail annule to tail tip; 1 long subventral pair (lateral to subventral) 3 annule widths posterior to last complete tail annule (just posterior to or 2 to 3 annule widths posterior to last complete tail annule) about equal in length to long setae in sub-latero-ventral rows on annulated tail region; 2 short pairs, 1 lateral pair (lateral to subventral) about 50%, and 1 subdorsal pair 2 annule widths posterior to last complete tail annule (present or absent, just posterior to or 1 to 2 annule widths posterior to last complete tail annule).

Females. —Similar to males. Ceph Acan-set about mid-rostrum, 5 to 7 annule widths or 9 to 11 μm anterior to first body annule. Two pairs of short, broad-based paravulval setae, 3 to 5 μm long, 1 pair anterior and 1 posterior to vulva. Longest SS on swollen esophageal region. Stout, broad-based setae in subventral rows just anterior to first SvAT, 2 to 6 on right and 2 to 4 on left side of body. Short setae intermingled with SlAT. SlAT with 11 to 25 long and 10 to 17 short tubes, not alternating. One of the SlAT adjacent to anus, and 1 posterior to anus. Caudal glands extend anterior to anus 1.8 to 2.6 times ABD. Anal flap short. One to 3 pairs of setae on non-annulated tail region; 1 subdorsal pair just posterior to last complete tail annule or about 25%; 1 lateral to subventral pair, present or absent, about 66%; 1 lateral to subdorsal pair, present or absent, about 50%.

Second-Stage Larvae. —Not observed.

Third-Stage Larvae. —L = 0.3 mm. Similar to adults. Amphids circular unispiral. CAT typical of genus. Ceph Acan-set absent. SlAT unpaired, 10 on right and 9 on left side of body, 1 of the SlAT posterior to anus. One subventral pair of setae 33% to 50% on non-annulated tail region, single dorsal seta about 33%.

Fourth-Stage Larvae. —L = 0.4 mm. Similar to adults. Amphids elongate loop-shaped. Six CAT, 1 sublateral and 2 subdorsal pairs in 2 transverse rows on rostrum. Ceph Acan-set 3 μm long, about mid-rostrum, 10 annule widths or 10 μm anterior to first body annule. Annules with very faint punctations, and a few scattered anteriorly directed, minute spines. Six stout setae in ventral row just anterior to first VAT. SlAT unpaired, 11 on right and 14 on left side of body. One of the SlAT on right and 2 on left side of body posterior to anus. Twelve VAT. Two pairs of setae on non-annulated tail region; 1 subdorsal pair 2 to 3 annule widths posterior to last complete tail annule or about 25%; and 1 lateral pair about 50%.

Holotype (♂).—Collected May 27, 1968 by I. M. Newell. Catalogue No. UCNC 1430, University of California, Davis.

Paratypes. —11 ♀♀, 25 ♂♂. Same data as holotype. Deposited at University of California, Davis (UCNC 1512 to 1513); USDA Nematode Collection, Beltsville, Maryland; U. S. National Museum of Natural History, Smithsonian Institution, Washington, D. C.; Station Marine D'Endoune et Centre

D'Océanographie, Marseille, France; and Laboratoria voor Morfologie en Systematiek, Museum voor Dierkunde, Gent, Belgium.

Type Habitat.—Marine, associated with coral in intertidal zone.

Type Locality.—Twenty miles north of Solano, Colombia.

Distribution.—Port Jackson, Australia; Solano, Colombia; and Isla Taboga, Panama.

Diagnosis.—Differs from other known species of *Dracograllus* by 1 pair of sublateral Ceph Acan-set on rostrum.

Third-stage larvae differ from third-stage *D. demani* n. sp. by greater number of SlAT, and absence of PS. Fourth-stage differ from other known fourth-stage larvae of *Dracograllus* by 1 pair of sublateral Ceph Acan-set on rostrum.

Dracograllus wieseri[24] n. sp.

(Figs. 143 to 145)

Measurements (1 ♀): L = 0.5 mm; b = 5.2; c = 7.4; V = 48%; CAT = 20 µm; No CAT = 8; SER (L/W) = 2.1; SER (E/L) = 21%; SS = 13-31 µm; first SlAT = 28 µm; last SlAT = 24 µm; first SvAT = 25 µm; last SvAT = 18 µm; No SlAT = 14; No SvAT = 12; Non-ann Term to Tail Length = 45%; T/ABD = 4.4.

(Holotype ♂): L = 0.6 mm; a = 10.5; b = 5.9; c = 8.4; CAT = 23 µm; No CAT = 8; SER (L/W) = 3.3; SER (E/L) = 24%; SS = 11-32 µm; first SlAT = 26 µm; last SlAT = 25 µm; first SvAT = 19 µm; last SvAT = 16 µm; No SlAT = 17; No SvAT = 13; Non-ann Term to Tail Length = 26%; T/ABD = 3.2; Spic = 46 µm; Gub = 19 µm.

Male Holotype.—Amphids elongate loop-shaped. Eight CAT, 2 sublateral and 2 subdorsal pairs in 2 transverse rows on rostrum, adjacent to amphids. Annulation faintly ornamented with granules and vacuoles to anus, most prominent on swollen esophageal region; tail annules without ornamentation. Longest SS on swollen esophageal and mid-body regions. Stout, broad-based, posteriorly directed setae in subventral rows just anterior to first SvAT, 6 on right and 5 on left side of body. Long and short setae intermingled with SlAT. SlAT with 6 long setae, not alternating with tubes, setae either equal to or shorter than adjacent tubes. SlAT with 3 long and 14 short tubes. All SlAT anterior to anus. Caudal glands extend anterior to anus 1.9 times ABD. Four pairs of stout, broad-based anal setae; 9 to 10 µm long, 2 anterior and 2 pairs posterior to anus. Anal flap absent. Two pairs of setae on non-annulated tail region, 1 subventral pair just posterior to last complete tail annule, 1 lateral pair slightly posterior and lateral to subventral pair.

Female.—Similar to male. Paravulval setae absent. Longest SS on swollen esophageal region. One pair of stout setae in subventral rows just anterior to first SvAT. Short setae intermingled with SlAT. SlAT with 3 long and 11 short tubes. One of the SlAT posterior to anus. Caudal glands extend anterior to anus 4.0 times ABD. Anal flap short. Two lateral pairs of setae on non-annulated tail region, position measured from last complete tail annule to tail tip; 1 just posterior to last complete tail annule, and 1 about 25%.

Second-, Third-, and Fourth-Stage Larvae.—Not observed.

Holotype (♂).—Collected July 18, 1966 by I. M. Newell. Catalogue No. UCNC 1431, University of California, Davis.

Paratype.—1 ♀. Same data as holotype. Deposited at University of California, Davis (UCNC 1514).

Type Habitat.—Marine, associated with green algae growing in tidal pool in high tide zone.

Type Locality.—Mas A Tierra Island, Juan Fernandez Islands, Chile.

Distribution.—Same as type locality.

Diagnosis.—Most closely resembles females of *D. chitwoodi* n. sp. differs by greater number of SlAT and SvAT. Differs from *D. gerlachi* n. sp., *D. stekhoveni* n. sp., and *D. falcatus* n. comb., by 8 CAT on rostrum; from *D. eira* n. comb. by CAT adjacent to amphids; and from *D. cobbi* n. sp., *D. mawsoni* n. sp., *D. demani* n. sp., and *D. kreisi* n. sp. by absence of PS. Differs from *D. solidus* n. comb. by 6 long setae intermingled with SlAT in males, elongate loop-shaped amphids and 1 SlAT posterior to anus in females; and from *D. timmi* n. sp. and *D. filipjevi* n. sp. by absence of cup-shaped cuticular collars on setae.

24. Named in honor of W. Wieser.

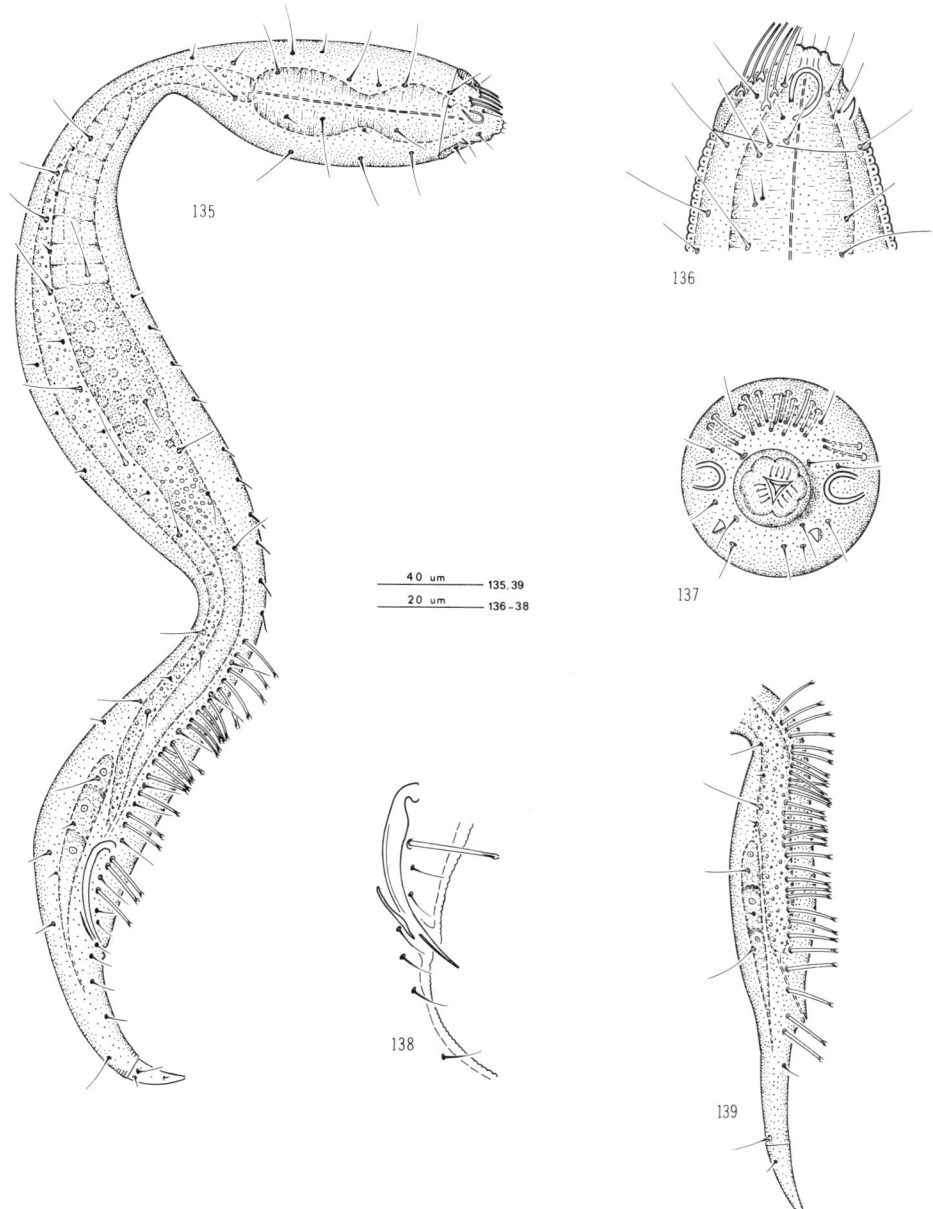

20. Figs. 135-139: *Dracograllus stekhoveni* n. sp. 135) Male, full length[a], SvAT and correct number of CAT not illustrated; 136) Male, head, correct number of CAT not illustrated; 137) Male, face view; 138) Male, anal region; 139) Female, posterior body region[a], SvAT not illustrated.

a. Short SS with SlAT not illustrated.

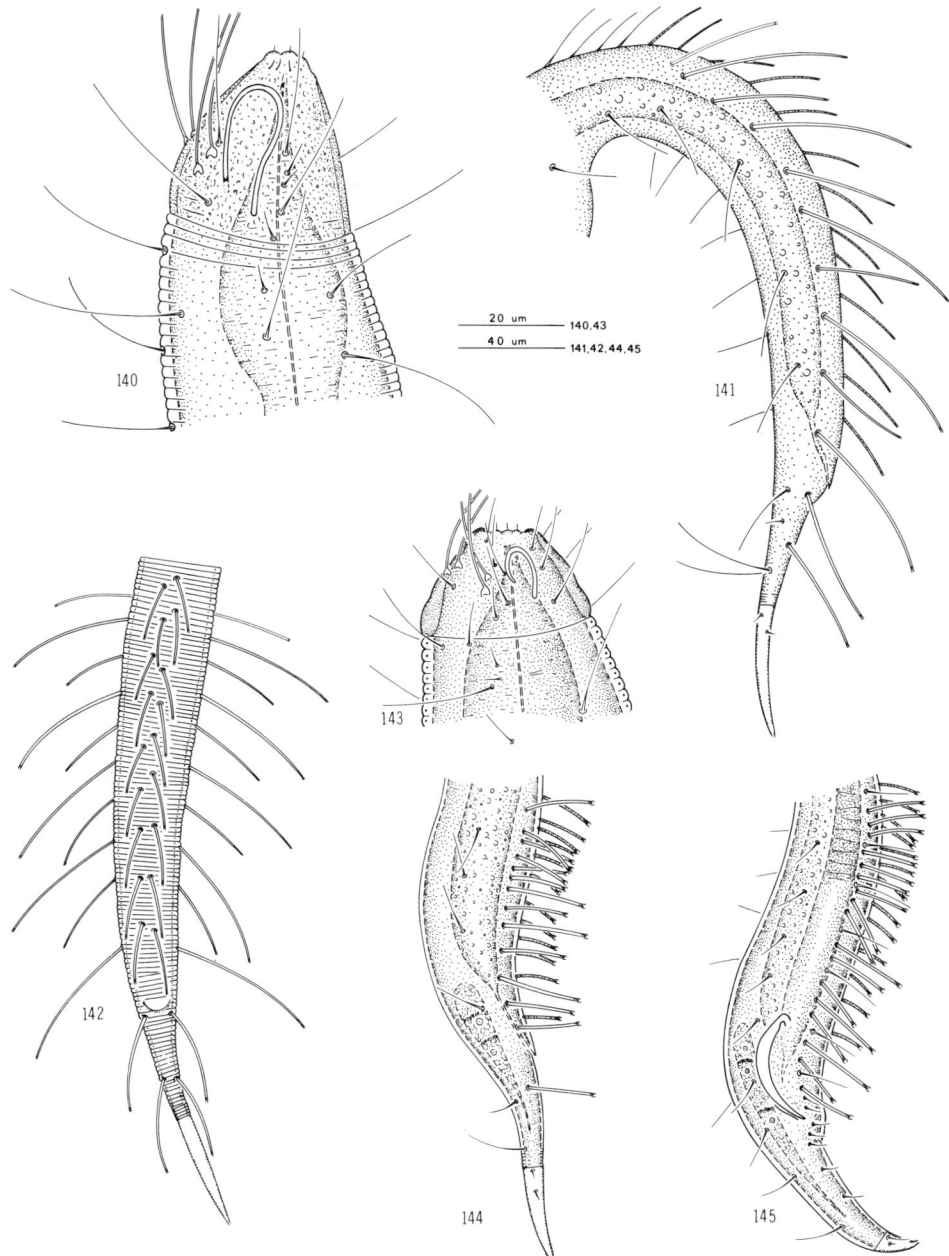

21. Figs. 140–142: *Dracotoranema trispinosum* n. gen. n. sp. 140) Male, head; 141) Female, posterior body region[a]; 142) Female, ventral view of PAT[b]. Figs. 143–145: *Dracograllus wieseri* n. sp. 143) Male, head; 144) Female, posterior body region[a]; 145) Male, posterior body region[a].

a. Short SS with SlAT not illustrated.
b. Total number and position of setae on non-annulated tail region not illustrated.

Dracograllus eira (Inglis, 1968) n. comb.

Syn: *Draconema eira* Inglis, 1968 n. syn.

Measurements (1 ♀): L = 0.6 mm; b = 4.6; c = 9.8; V = 56%; CAT = 15 μm; No CAT = 8; SER (L/W) = 2.9; SER (E/L) = 30%; SS = 17-31 μm; first SlAT = 18 μm; last SlAT = 29 μm; SvAT = 16 μm; No SlAT = 12-14; No SvAT = 8; Non-ann Term to Tail Length = 41%; T/ABD = 2.4.

(4 ♂♂): L = 0.5 mm; a = 7.4 (7.2-7.5); b = 4.4 (4.3-4.5); c = 9.5 (9.4-9.5); CAT = 12 (13-14) μm; No CAT = 8; SER (L/W) = 2.9; SER (E/L) = 30%; SS = 14-33 μm; first SlAT = 18 μm; last SlAT = 28 μm; SvAT = 16 μm; No SlAT = 12-14; No SvAT = 8; Non-ann Term to Tail Length = 48%; T/ABD = 2.4 (2.0-2.8); Spic = 49 (48-51) μm; Gub = 12 (12-13) μm.

Males Emended.—Amphids elongate loop-shaped. Eight CAT, 2 sublateral and 2 subdorsal pairs in 2 transverse rows on rostrum, anterior to amphids. Annulation without ornamentation. Longest SS on mid-body region. Seven to 9 long, stout, posteriorly directed setae in ventral row just anterior to first SvAT. Long and short setae intermingled with SlAT. Four long setae with SlAT, not alternating with tubes, setae shorter than adjacent tubes. SlAT with long and short tubes, longest tubes posterior to anus. Two of SlAT posterior to anus, posterior tube on non-annulated tail region. Caudal glands present. Three pairs of anal setae, 2 pairs anterior and 1 posterior to anus. Anal flap absent. Three pairs of setae on non-annulated tail region, position measured from last complete tail annule to tail tip; 1 long subdorsal pair just posterior to last complete tail annule, about equal in length to long setae on mid-body region; 1 short subventral pair just posterior to last complete tail annule; 1 short lateral pair about 33%.

Female Emended.—Similar to males. Paravulval setae absent. Longest SS on mid-body region. Seven to 9 long, stout, setae in ventral row just anterior to first SvAT. Short setae intermingled with SlAT. SlAT with long and short tubes, longest tubes posterior to anus. Two of the SlAT posterior to anus, posterior tube on non-annulated tail region. One subdorsal pair of setae on non-annulated tail region about 25%.

Larval Stages.—Not observed.

Type Specimens (1 ♀, 4 ♂♂).—Collected September 12, 1961. Catalogue Nos. 1111, 1112, 1113, and 1114, British Museum (Natural History), London, England.

Type Habitat.—Marine, in intertidal zone.

Type Locality.—St. Vincent's Bay, New Caledonia.

Distribution.—Same as type locality.

Diagnosis.—Differs from all other known species of *Dracograllus* by all CAT anterior to amphids, and 1 SlAT on non-annulated tail region. Males differ from all other known *Dracograllus* by some SlAT posterior to anus.

Type specimens were available for study, and from these specimens this species is easily distinguished from other species of *Dracograllus*. After comparing the type specimens with the original description it must be noted that the total CAT on the rostrum is 8 instead of 6 as indicated in the original description.

Dracograllus falcatus (Irwin-Smith, 1918) n. comb.

Syn: *Chaetosoma falcatum* Irwin-Smith, 1918 n. syn.
  Notochaetosoma falcatum (Irwin-Smith, 1918) Cobb, 1929 n. syn.
  Drepanonema falcatum (Irwin-Smith, 1918) Cobb, 1933 n. syn.
  Claparediella falcatum (Irwin-Smith, 1918) Filipjev, 1934 n. syn.
  Draconema falcatum (Irwin-Smith, 1918) Kreis, 1938 n. syn.
  Tristicochaeta falcata (Irwin-Smith, 1918) Johnston, 1938 n. syn.

Measurements (1 ♀): L = 0.9 mm; b = 11.0; c = 11.0; V = 43%; CAT = 15 μm; No CAT = 12; SER (L/W) = 2.2; SER (E/L) = 13%; SS = 18-22 μm; SlAT = 18-22 μm; SvAT = 18-22 μm; No SlAT = 21; No SvAT = 23; Non-ann Term to Tail Length = 48%; T/ABD = 2.5.

(1 ♂): L = 0.8 mm; a = 10.8; b = 10.3; c = 10.3; CAT = 15 μm; No CAT = 12; SER (L/W) = 1.9; SER (E/L) = 14%; SS = 1-26 μm; SlAT = 18-22 μm; SvAT = 18-22 μm; No SlAT = 12; No SvAT = 17; Non-ann Term to Tail Length = 32%; T/ABD = 2.3; Spic = 71 μm; Gub = 12 μm.

Male Emended.—Amphids elongate loop-shaped. Twelve CAT, 3 sublateral and 3 subdorsal pairs in 3 transverse rows on rostrum; anterior row anterior to, middle row adjacent to and posterior row posterior to amphids. Annulation without ornamentation. Longest SS on swollen esophageal and midbody regions. Three to 4 short, stout, setae in subventral rows just anterior to first SvAT. Seven long setae intermingled with SlAT, setae about equal in length to adjacent tubes. All SlAT anterior to anus. Caudal glands extend anterior to anus 1.3 times ABD. Anal setae and anal flap absent. Three pairs of setae on non-annulated tail region, position measured from last complete tail annule to tail tip; 1 subdorsal pair just posterior to last complete tail annule; 2 lateral pairs, 1 about 33%, and 1 about 50%.

Female Emended.—Similar to male. Two pairs of paravulval setae, 1 anterior and 1 pair posterior to vulva. Three of the SlAT posterior to anus. One subdorsal pair of setae on non-annulated tail region just posterior to last complete tail annule.

Larval Stages.—Larvae were observed by Irwin-Smith in 1918. She stated that the number of posterior adhesion tubes increases with age and observed 12 SlAT on the youngest specimen.

Type Specimens (1 ♀, 1 ♂).—Collected from 1915 to 1918 by V. A. Irwin-Smith. Catalogue Nos. W452 and W453, Australian Museum, Sydney, New South Wales, Australia.

Type Habitat.—Marine, associated with seaweed and shells at 1.2 to 1.5 meters below low tide zone.

Type Locality.—Cremorne, Port Jackson, New South Wales, Australia.

Distribution.—Cremorne, Port Jackson; Long Reef; and Vaucluse, Australia.

Diagnosis.—Most closely resembles *D. gerlachi* n. sp. differs by shorter swollen esophageal region, 3 to 4 stout setae in subventral rows just anterior to first SvAT, and longer spicules in males; fewer SlAT and 3 SlAT posterior to anus in females. Differs from *D. stekhoveni* n. sp. by absence of sublateral Ceph Acan-set on rostrum, and from other known species of *Dracograllus* by 12 CAT on rostrum.

The type specimens were available for study. Although these specimens were in very poor condition, we were able to use these specimens plus the original description to distinguish this species from other species of *Dracograllus*.

Discussion.—T. H. Johnston in 1938, in the same publication in which he proposed the new family Drepanonematidae, made the new combination *Tristicochaeta falcatum* (Irwin-Smith, 1918). (For further details, see Taxonomic History.)

Dracograllus solidus (Gerlach, 1952) n. comb.

Syn: *Draconema solidum* Gerlach, 1952 n. syn.

Measurements (3 ♀♀): L = 0.7 (0.6–0.8) mm; b = 4.9–5.5; c = 8.7–10.0; V = ?; CAT = 17 μm; No CAT = 8; SER (L/W) = 2.4; SER (E/L) = 23%; SS = 20–50 μm; SlAT = 33 μm; SvAT = 21 μm; No SlAT = 8–11; No SvAT = 8–11; Non-ann Term to Tail Length = 46%; T/ABD = 2.7–2.9.

(1 ♂): L = 0.7 mm; a = 9.6; b = 5.5; c = 8.8; CAT = 18 μm; No CAT = 8; SER (L/W) = 2.4; SER (E/L) = 23%; SS = 20–50 μm; SlAT = 22 μm; SvAT = 18 μm; No SlAT = 7; No SvAT = 11; Non-ann Term to Tail Length = 29%; T/ABD = 2.6; Spic = 46 μm; Gub = 20 μm.

Male.—Amphids elongate loop-shaped. Eight CAT, 2 sublateral and 2 subdorsal pairs in 2 transverse rows on rostrum, adjacent to amphids. Annulation without ornamentation. Longest SS on swollen esophageal region. Six stout setae in subventral rows just anterior to first SvAT. Eleven long setae intermingled with SlAT, 1 seta between first and second tubes and 2 setae between each remaining tube, setae about equal to length of adjacent tubes. All SlAT anterior to anus. Caudal glands present. Anal setae and anal flap absent. Three pairs of setae on non-annulated tail region, position measured from last complete tail annule to tail tip; 2 latero-subventral pairs about 33% and 50%; 1 latero-subdorsal pair about 25%.

Females.—Similar to male. Amphids elongate unispiral. Paravulval setae absent. Two of the SlAT posterior to anus. One subventral pair of setae on non-annulated tail region about 50%.

Larval Stages.—Not observed.

Type Habitat.—Marine.

Type Locality.—Banyuls, France; on the Mediterranean Sea.

Distribution.—Same as type locality.

This species has also been recorded from the Bay of Biscay, and the Mascarene Islands.

Diagnosis.—Most closely resembles *D. wieseri* n. sp. and *D. chitwoodi* n. sp. differs by 11 long setae intermingled with SlAT in males; unispiral amphids, and 2 SlAT posterior to anus in females. Differs from *D. filipjevi* n. sp. and *D. timmi* n. sp. by absence of setae with cup-shaped cuticular collars; and from *D. eira* n. comb. by CAT adjacent to amphids. Differs from *D. cobbi* n. sp., *D. mawsoni* n. sp., *D. demani* n. sp., and *D. kreisi* n. sp. by absence of PS; and from *D. stekhoveni* n. sp., *D. gerlachi* n. sp., and *D. falcatus* n. comb. by 8 CAT on rostrum.

Type material of this species was not available for study. This species is easily distinguishable from other known species of *Dracograllus* from the adequate original description.

Genus *Dracotoranema*[25] n. gen.

Diagnosis: Draconematinae. Nematodes 0.6 to 0.8 mm long. Body shape as in *Dracograllus*. Swollen esophageal region generally longer than in *Draconema* and *Paradraconema*, averaging 23% of total body length. Rostrum ornamented with subsurface markings, setae present. Amphids very large, conspicuous, slightly dorsal on rostrum. Male amphids elongate loop-shaped, ventral arm usually longer than dorsal arm. Female amphids elongate unispiral. Eight CAT, 2 sublateral and 2 subdorsal pairs in 2 transverse rows on rostrum, adjacent to amphids. Body annules ornamented to anus, tail annules without ornamentation. Females without paravulval setae. Male SlAT all anterior to anus. Females with some of the SlAT posterior to anus. SlAT with conspicuously long and short, slender tubes, differences in length obvious, long and short tubes usually alternating. Males with uniformly tapered anal setae. Males with 1 pair of large subventral preanal Corn-set, and a single large ventral Corn-set just anterior to first SvAT. Anal flap absent. Setae on non-annulated tail region.

Type Species: *Dracotoranema trispinosum* n. sp.

Differs from other known genera in Draconematinae by conspicuously long and short slender SlAT, very large amphids; a single large ventral Corn-set just anterior to first SvAT in males. Differs from *Draconema* Cobb, 1913 by absence of prominently enlarged annules on swollen esophageal region; presence of Corn-set in males; and SlAT posterior to anus in females. Differs from *Paradraconema* n. gen. by absence of eyespots and sublateral Ceph Acan-set on rostrum, and from *Dracograllus* n. gen. by absence of PS, and presence of Corn-set in males.

Larval Stages: No first- or second-stage larvae available in *Dracotoranema*. Third-stage larvae 0.4 to 0.5 mm long. Similar to adults. Esophageal region swollen, mid-body region only slightly swollen. Length of swollen esophageal region similar to third-stage *Dracograllus*. Amphids large, conspicuous, elongate unispiral. Three CAT, 2 subdorsals and 1 dorsal in 2 transverse rows on rostrum. Six rows of SS on swollen esophageal region, 4 sublaterals, 1 dorsal and 1 ventral. Five rows of SS on mid-body region, 4 sublateral and 1 dorsal; 4 sublateral rows on annulated tail region. Some rows with alternating long and short setae. SlAT in 2 longitudinal rows; long and short tubes present, usually alternating. All SlAT anterior to anus, last tube just anterior to anus. Anal flap absent. Differentiated from other stages by 3 CAT and SlAT in 2 rows.

Fourth-stage larvae 0.6 to 0.7 mm long. Similar to adults. Esophageal region and mid-body region swollen, especially in young females. Length of swollen esophageal region similar to fourth-stage *Dracograllus*. Amphids large, conspicuous, elongate unispiral. Four CAT, 2 sublaterals and 2 subdorsals in 2 transverse rows on rostrum. Seven rows of SS on swollen esophageal region, 4 sublaterals, 2 subdorsals and 1 ventral. Six rows of SS on mid-body region, 4 sublaterals, 1 dorsal and 1 ventral; 4 sublateral rows on annulated tail region. Some rows of setae with alternating long and short setae, some setae with cuticular collars. Three longitudinal rows of PAT, 2 sublateral rows and 1 ventral. Anal flap absent. Differentiated from other stages by 4 CAT, and PAT in 3 rows.

---

25. *Dracotoranema*, Gr. n., dragon nematode with projections.

Dracotoranema trispinosum n. sp.
(Figs. 140 to 142, 146)

Measurements (8 ♀♀): L = 0.7 (0.6–0.8) mm; b = 6.1 (5.0–7.9); c = 6.5 (5.7–7.3); V = 44 (38–47)%; CAT = 27 (23–31) μm; SER (L/W) = 3.0 (2.5–3.8); SER (E/L) = 23 (20–27)%; SS = 19–54 μm; first SlAT = 37 (32–49) μm; last SlAT = 45 (40–48) μm; first SvAT = 29 (26–31) μm; last SvAT = 33 (29–38) μm; No SlAT = 13 (12–13); No SvAT = 9 (8–13); Non-ann Term to Tail Length = 57 (52–61)%; T/ABD = 5.3 (4.8–6.1).

(4 ♂♂): L = 0.7 mm; a = 11.4 (10.8–12.1); b = 6.1 (5.7–6.6); c = 8.6 (7.4–9.4); CAT = 27 (25–30) μm; SER (L/W) = 2.9 (2.8–3.0); SER (E/L) = 23 (22–24)%; SS = 21–66 μm; first SlAT = 39 (30–47) μm; last SlAT = 65 (44–77) μm; first SvAT = 26 (24–27) μm; last SvAT = 32 (28–35) μm; No SlAT = 10; No SvAT = 6 (6–7); Non-ann Term to Tail Length = 26 (26–27)%; T/ABD = 3.4 (3.1–3.5); Spic = 61 (59–64) μm; Gub = 18 (17–19) μm; Ventral Corn-set = 15 (15–17) μm; Preanal Corn-set = 17 (16–18) μm.

(Holotype ♂): L = 0.7 mm; a = 11.6; b = 6.0; c = 9.0; CAT = 30 μm; SER (L/W) = 2.8; SER (E/L) = 22%; SS = 23–45 μm; first SlAT = 47 μm; last SlAT = 73 μm; first SvAT = 27 μm; last SvAT = 35 μm; No SlAT = 10; No SvAT = 7; Non-ann Term to Tail Length = 26%; T/ABD = 3.4; Spic = 61 μm; Gub = 19 μm; Ventral Corn-set = 15 μm; Preanal Corn-set = 18 μm.

Male Holotype.—(Figures in parentheses refer to range within species.) Amphids elongate loop-shaped (some specimens ventral arm appearing to form unispiral), very large, conspicuous, almost as long as rostrum. CAT typical of genus. Annules ornamented with 2 transverse rows of dot-like punctations and subsurface granulation to anus. Tail annules without ornamentation. Longest SS on swollen esophageal region (some specimens longest on both swollen esophageal and mid-body regions). Some setae on swollen esophageal region with inconspicuous cuticular collars. Short setae alternate with SlAT. One long seta with SlAT, seta shorter than adjacent tubes. SlAT with 6 long and 4 short tubes (4 to 6 long and 4 to 6 short tubes). Long and short SlAT alternate (rarely with 2 long tubes or 2 short tubes together). Length differences between long and short SlAT conspicuous, averaging 15 μm (5 to 29 μm). All SlAT anterior to anus. Caudal glands extend anterior to anus 2.8 (2.2 to 2.8) times ABD. Single large ventral Corn-set just anterior to first SvAT, and 1 pair of large preanal subventral Corn-set just posterior to last SvAT. Three pairs of anal setae, 13 to 14 μm (13 to 16 μm) long, 1 adjacent to and 2 pairs posterior to anus, setae close together. Anal region swollen, especially posterior to anus (fig. 146). One long pair of setae on non-annulated tail region subventral on right and lateral on left side of body, about 33%, position measured from last complete tail annule to tail tip, about equal in length to $1/2$ ABD; short single dorsal seta about 50% (present or absent).

Females.—Similar to males. Amphids elongate unispiral, very large, conspicuous, almost as long as rostrum. Longest SS on swollen esophageal and mid-body regions. Short setae usually alternating with SlAT. SlAT with 5 to 6 long and 7 to 8 short tubes, usually alternating. Length differences between long and short SlAT conspicuous, averaging 12 to 27 μm. One to 2, usually 2, of the SlAT posterior to anus. Caudal glands extend anterior to anus 1.3 to 2.6 times ABD. Anal region swollen. Two to 4 pairs of setae on non-annulated tail region; 1 subventral pair about 33%; 1 lateral to subdorsal pair, present or absent, just posterior to or on last complete tail annule; 1 subdorsal pair, present or absent, about 50%; 1 subdorsal pair about 13%.

Third-Stage Larvae.—L = 0.4 to 0.5 mm. Similar to adults. CAT typical of genus. Amphids elongate unispiral, very large, conspicuous, almost as long as rostrum. Four to 5 pairs of SlAT, all SlAT anterior to anus. Anal region slightly swollen. Usually 1 long lateral pair of setae, about equal in length to ABD, about 50% on non-annulated tail region.

Fourth-Stage Larvae.—L = 0.6 to 0.7 mm. Similar to adults. CAT typical of genus. Amphids elongate unispiral, very large, conspicuous, almost as long as rostrum. Some SS on swollen esophageal region with cuticular collars. Six to 7 pairs of SlAT, 1 of the SlAT posterior to anus. Six to 7 VAT; posterior 2 to 4 tubes slightly subventral, almost forming 2 longitudinal rows. Two to 4 pairs of setae on non-annulated tail region; 1 lateral to subdorsal pair just posterior to last complete tail annule or 5 annule widths posterior to last tail annule; 1 subventral pair 13% to 25%; 2 subdorsal pairs present or absent, 1 just posterior to last complete tail annule, and 1 about 50%.

Holotype (♂).—Collected February 26, 1972 by P. Vitiello. Catalogue No. UCNC 1432, University of California, Davis.

Paratypes. —3 ♀♀, 3 ♂♂. Same data as holotype. Deposited at University of California, Davis (UCNC 1515 to 1516); and USDA Nematode Collection, Beltsville, Maryland.

Larval Stages. —5 third, 2 fourth. Same data as holotype. Deposited in same nematode collections as paratypes.

Type Habitat. —Marine, at 20 meters.

Type Locality. —Southwest of the Pomègues Ratonneau jetty, near Marseille, France.

Distribution. —Region of Marseille, France.

Diagnosis. —Differs by swollen anal region, number of SlAT (♀♀ = 8 to 13; ♂♂ = 10) and SvAT (♀♀ = 8 to 13; ♂♂ = 6 to 7), and some setae on swollen esophageal region with cuticular collars. Males differ by 3 large Corn-set, a single ventral mid-body seta and 1 preanal pair.

Third-stage larvae differ by 4 to 5 pairs of SlAT, and 1 long lateral pair of setae on non-annulated tail region about 50%. Fourth-stage differ by swollen anal region, 6 to 7 pairs of SlAT, 1 SlAT posterior to anus, 6 to 7 VAT, and some setae on swollen esophageal region with cuticular collars.

Genus *Prochaetosoma* Micoletzky, 1922

Diagnosis Emended: Prochaetosomatinae. Nematodes 0.5 to 1.0 mm long. Most relaxed specimens assume an open "S" shape. Esophageal region only slightly swollen, usually distinctly arched ventrally (fig. 151). Greatest body width at mid-body in females, usually at esophageal and at PAT region in males. Rostrum broadly flattened anteriorly (fig. 148), with or without ornamentation, setae present. Amphids conspicuous, dorso-lateral on rostrum; elongate loop-shaped, loop-shaped (horseshoe-shaped), modified loop appearing circular, circular or elongate unispiral, or doubled spiral. Four to 14 CAT, paired or unpaired; in 1 to 3 transverse rows usually posterior to rostrum, some species with tubes both on and posterior to rostrum. Posterior row 1 or less rostral widths or 4 to 11 annules posterior to rostrum. Buccal cavity moderately developed with conspicuous dorsal tooth, with or without 1 to 2 minute ventral teeth. Posterior esophageal bulb with cuticularized valve. Cuticle not unusually thick. Annules with or without subcuticular markings, dot-like punctations, or annular ridges with spine-like projections. SS usually in distinct rows on body, some rows with alternating long and short setae. With or without stout ventral setae just anterior to SvAT or stout ventral conically shaped broad-based setae just posterior to SvAT. Paravulval setae usually present. Four longitudinal rows of PAT; except in *P. vitielloi* n. sp. with 4 anterior rows (2 sublateral and 2 subventral) and 3 posterior rows (2 sublateral and 1 ventral). Number of PAT variable between species. With or without last SvAT greatly reduced in size. Males usually with uniformly tapered anal setae. Anal flap present or absent. Tails cylindrical-conoid. Spinneret usually present; except, appears lacking in *P. vitielloi* n. sp. and *P. mediterranicum* n. sp. Setae present or absent on non-annulated tail region.

Type Species: *Prochaetosoma primitiva* (Steiner, 1916) Micoletzky, 1922

Differs from other known genera in Prochaetosomatinae by moderately developed buccal cavity with conspicuous dorsal tooth, and posterior esophageal bulb with cuticularized valve.

Larval Stages: No first-stage larvae available in *Prochaetosoma*. Second-stage larvae 0.3 to 0.4 mm long. Similar to adults. Widest part of body in esophageal region. Rostral setae absent. Amphids conspicuous, dorso-lateral on rostrum; loop-shaped, modified loop appearing circular, unispiral, or doubled spiral. Single dorsal CAT, posterior to rostrum. Buccal cavity moderately developed with dorsal tooth. If adults have minute ventral teeth, these teeth present or absent. A few SS in distinct rows. Some rows with alternating long and short setae. Two pairs of SlAT in 2 longitudinal rows. Anal flap present or absent. Setae present or absent on non-annulated tail region. Differentiated from other 2 stages by 1 CAT, 2 pairs of SlAT in 2 rows.

Third-stage larvae 0.3 to 0.5 mm long. Similar to adults. Widest part of body in esophageal region. Rostral setae present or absent. Amphids conspicuous, dorso-lateral on rostrum, modified loop appearing circular, or doubled spiral. Three CAT, 2 subdorsal and 1 dorsal, in 1 transverse row posterior to rostrum. Buccal cavity moderately developed with dorsal tooth. If adults have minute ventral teeth, these teeth present but obscure. SS in distinct rows, some rows with alternating long and short setae. Three pairs of SlAT in 2 longitudinal rows. Anal flap present. Setae present on non-annulated tail region. Differentiated from other 2 stages by 3 CAT, 3 pairs of SlAT in 2 rows.

Fourth-stage larvae 0.4 to 0.8 mm long. Similar to adults. Widest part of body in esophageal region. Rostral setae present or absent. Amphids conspicuous, dorso-lateral on rostrum, modified loop

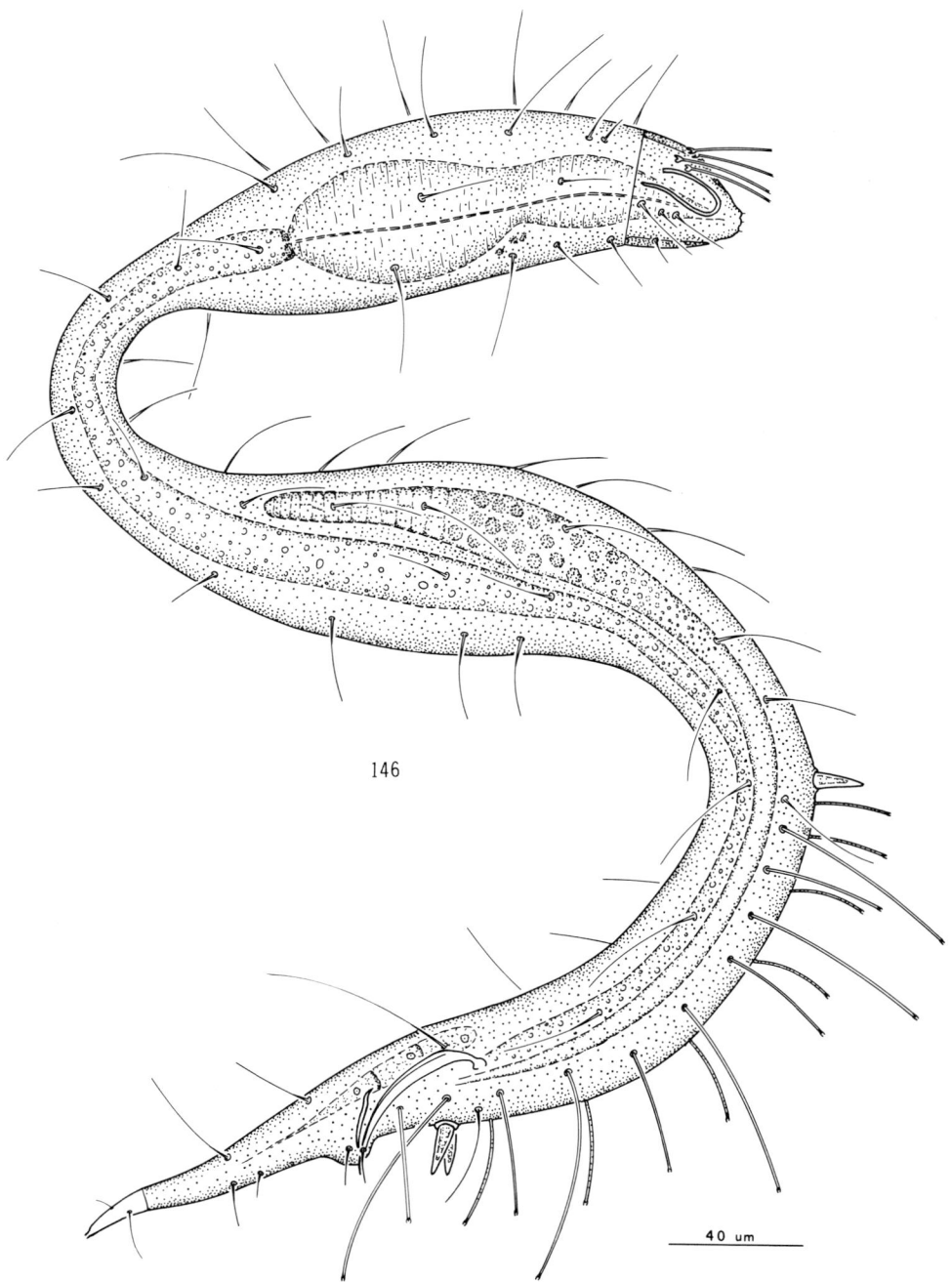

22. Fig. 146: *Dracotoranema trispinosum* n. sp. 146) Male, full length[a], note ventral and preanal Corn-set.
   a. Short SS with SlAT not illustrated.

appearing circular, unispiral, or doubled spiral. Four to 8 CAT, in 1 or 2 transverse rows, usually posterior to rostrum. Buccal cavity moderately developed with dorsal tooth. If adults have minute ventral teeth, these teeth present. SS in distinct rows, some rows with alternating long and short setae. PAT in 3 longitudinal rows, 2 sublateral and 1 ventral, number of tubes variable between species. Anal flap present or absent. Setae present on non-annulated tail region. Differentiated from other 2 stages by 4 to 8 CAT, 3 rows of PAT.

Discussion.—G. Steiner in 1916, described *Chaetosoma primitivum* from the Barents Sea, with a young specimen which was a fourth-stage larva. From the distinctive taxonomic characters of this species, H. Micoletzky in 1922 proposed the new genus *Prochaetosoma* with *Chaetosoma primitivum* Steiner, as the type species. Thus, Micoletzky's name *Prochaetosoma* invalidates the *Prochaetosoma* of Baylis and Daubney, 1926.

Because of the stable morphological characters in fourth-stage larvae of the superfamily and the distinctive taxonomic characters given in the original description of this specimen, we conclude that *P. primitiva* is a good species.

Key to Species of *Prochaetosoma* (Adults)

1. PAT with 4 longitudinal rows, 2 sublaterals and 2 subventrals *2*
   PAT with 4 anterior longitudinal rows, 2 sublaterals and 2 subventrals; and 3 posterior rows, 2 sublaterals and 1 ventral (fig. 154) . . . . *vitielloi* n. sp. (p. 96)
2. Females with 15 SlAT[26]; 5 ventral stout setae just anterior to SvAT, with distal ends directed posteriorly. Males with 18 SlAT *3*
   Females with 8 to 12 SlAT; without ventral stout setae just anterior to SvAT. Males with 5 to 8 SlAT *4*
3. Females with 15 SlAT, 4 to 5 SvAT . . . . *lugubre* (Gerlach, 1957) n. comb. (p. 105)
   Males with 18 SlAT, 9 SvAT . . . . *longicapitata* (Allgén, 1932) n. comb. (p. 104)
4. Eight or less CAT (fig. 152) *5*
   Fourteen CAT (fig. 149) . . . . *cayense* n. sp. (p. 94)
5. Males with 5 to 8 SlAT. Females with 8 to 9 SlAT *6*
   Females with 12 SlAT . . . . *arcticum* (Kreis, 1963) n. comb. (p. 101)
6. Males and females with 2 to 3 SvAT . . . . *mediterranicum* n. sp. (p. 100)
   Males with 7 to 8 SvAT. Females with 10 to 11 SvAT . . . .
   . . . . *campbelli* (Allgén, 1932) n. comb. (p. 103)

Key to Species of *Prochaetosoma* (Fourth-Stage Larvae)

1. Five pairs of SlAT; 2 to 9 VAT *2*
   Six pairs of SlAT; 13 VAT . . . . . *primitiva* (Steiner, 1916) Micoletzky, 1922 (p. 93)
2. Nine VAT *3*
   Two to 3 VAT *4*
3. Amphids double spiral, ³⁄₄ spiral doubled . . . . *cayense* n. sp. (p. 94)
   Amphids unispiral . . . . *campbelli* (Allgén, 1932) n. comb. (p. 103)
4. Three VAT . . . . *vitielloi* n. sp. (p. 96)
   Two VAT . . . . *mediterranicum* n. sp. (p. 100)

Prochaetosoma primitiva (Steiner, 1916) Micoletzky, 1922

Syn: *Chaetosoma primitivum* Steiner, 1916
    Draconema primitivum (Steiner, 1916) Filipjev, 1918
    Drepanonema primitivum (Steiner, 1916) Cobb, 1933 n. syn.
    Claparediella primitivum (Steiner, 1916) Filipjev, 1934 n. syn.

26. All counts and measurements on right side of body.

Measurements (1 Fourth-Stage Larva): L = 0.5 mm; b = 5.2; c = 8.4; No CAT = 4; SlAT = 28-29 μm; VAT = 18-25 μm; No SlAT = 6; No VAT = 13; Non-ann Term to Tail Length = 51%; T/ABD = 3.5.

Males, Females, Second- and Third-Stage Larvae.—Not observed.

Fourth-Stage Larva. Rostrum with setae. Amphids unispiral, about $1/8$ spiral doubled. Four CAT, in 1 transverse row, less than 1 rostral width or about 11 annules posterior to rostrum. Buccal cavity with small dorsal tooth. Annulation without ornamentation. SS in 6 rows on body, 4 sublateral, 1 dorsal and 1 ventral. Tail without setae. Six SlAT and 13 VAT. Caudal glands and spinneret present.

Type Habitat.—Marine.
Type Locality.—Barents Sea.
Distribution.—Same as type locality.
Diagnosis.—Differs from other known fourth-stage *Prochaetosoma* by 13 VAT.

Type material of this species was not available for study. (See discussion following generic description.)

Prochaetosoma cayense[27] n. sp.

(Figs. 147 to 150)

Measurements (19 ♀♀): L = 0.9 (0.6-1.0) mm; b = 9.8 (7.4-12.0); c = 9.8 (7.4-12.0); V = 51 (50-55)%; CAT = 26 (21-30) μm; No CAT = 14; SS = 8-64 μm; first SlAT = 39 (33-45) μm; last SlAT = 21 (18-26) μm; first SvAT = 30 (23-37) μm; last SvAT = 16 (13-20) μm; No SlAT = 11 (10-12); No SvAT = 16 (14-18); Non-ann Term to Tail Length = 65 (61-68)%; T/ABD = 5.3 (4.5-7.2).

(7 ♂♂): L = 0.9 (0.8-1.0) mm; a = 19.4 (16.9-22.2); b = 8.9 (7.8-10.0); c = 8.0 (6.7-10.0); CAT = 26 (22-29) μm; No CAT = 14; SS = 7-58 μm; first SlAT = 40 (36-44) μm; last SlAT = 24 (21-28) μm; first SvAT = 31 (28-35) μm; last SvAT = 14 (9-17) μm; No SlAT = 8 (7-8); No SvAT = 15 (14-17); Non-ann Term to Tail Length = 54 (49-60)%; T/ABD = 4.5 (4.1-5.0); Spic = 59 (54-63) μm; Gub = 16 (11-19) μm.

(Holotype ♂): L = 0.8 mm; a = 17.7; b = 7.8; c = 7.8; CAT = 29 μm; No CAT = 14; SS = 10-56 μm; first SlAT = 40 μm; last SlAT = 28 μm; first SvAT = 32 μm; last SvAT = 17 μm; No SlAT = 8; No SvAT = 15; Non-ann Term to Tail Length = 53%; T/ABD = 4.4; Spic = 54 μm; Gub = 15 μm.

(4 Fourth-Stage Larvae): L = 0.5-0.8 mm; b = 6.5-11.3; c = 6.8-10.0; No CAT = 8; SlAT = 24-34 μm; VAT = 14-30 μm; No SlAT = 5; No VAT = 9; Non-ann Term to Tail Length = 54-71%; T/ABD = 4.5-5.1.

Male Holotype.—(Figures in parentheses refer to range within species.) Rostrum with subcuticular markings of granules and vacuoles, and deep punctations. Amphids loop-shaped, horseshoe-shaped. Fourteen CAT (fig. 149), unpaired; in 3 transverse rows posterior to rostrum, posterior row less than 1 rostral width or 4 (4 to 5) annules posterior to rostrum. Buccal cavity with dorsal tooth, 1 minute subventral tooth slightly posterior to dorsal tooth (fig. 148). Annulation with obscure subcuticular granulation, most prominent on tail annules. Annules in esophageal region with obscure spine-like projections appearing as dot-like punctations on anterior margins of annular ridges, similar to fig. 166; annules slightly larger than remaining annules, anterior margins directed anteriorly. Longest SS on mid-body region (some specimens longest occur on both esophageal and mid-body regions). SS on esophageal region not in distinct rows, probably 8, 4 sublateral, 2 subdorsal and 2 subventral. Eight rows of SS on mid-body, 4 sublateral, 2 subdorsal and 2 subventral; 4 sublateral rows on annulated tail region. Four longitudinal rows of PAT, 2 sublateral and 2 subventral. Long and short setae intermingled with SlAT. Four long setae with SlAT, not alternating with tubes, setae longer than adjacent tubes. Last SvAT distinctly shorter than adjacent tube on left side of body (usually on only 1 side of body, sometimes last SvAT shorter on both sides). Difference between shorter last SvAT and adjacent tube 7 μm (6 to 8 μm). Caudal glands extend anterior to

27. *cayense*, small island or key.

anus 1.8 (1.8–2.4) times ABD. Spinneret present. Two pairs of anal setae, 12 to 13 μm (8 to 14 μm) long, both pairs anterior to anus. Anal flap short. Seven pairs of setae on non-annulated tail region, position measured from last complete tail annule to tail tip; 2 subdorsal pairs, 1 long pair about 25% equal in length to long setae in subdorsal rows opposite PAT, 1 short pair just posterior to long pair; 3 short subventral pairs, 1 just posterior to last complete tail annule, 1 about 33%, and 1 just posterior to second pair; 2 short latero-subdorsal pairs, 1 about 75%, and 1 just posterior to first pair.

Females.—Similar to males. Amphids doubled spiral. CAT posterior row less than 1 rostral width or 4 to 6 annules posterior to rostrum. Two subventral pairs of anteriorly directed paravulval setae, 12 to 17 μm long, 1 pair anterior and 1 posterior to vulva. Longest SS on esophageal and mid-body regions. Short setae intermingled with SlAT. Last SvAT not distinctly shorter than adjacent tube. Caudal glands extend anterior to anus 2.0 to 3.0 times ABD. Anal flap long, faintly crenate. Five to 6 pairs of setae on non-annulated tail region; 1 short subventral to sublateral, usually sublateral, pair about 50%; 4 to 5 subdorsal pairs, 1 long pair about 33% and equal in length to long setae in subdorsal rows opposite PAT, 1 short pair (present or absent) just posterior to or on last complete tail annule, 1 short pair just posterior to second pair, 1 short pair just posterior to third pair, and 1 short pair about 75%.

Second-Stage Larvae.—L = 0.3 mm. Similar to adults. Rostrum ornamentation obscure, setae absent. Amphids doubled spiral, about $1/2$ spiral doubled. CAT typical of genus, tube less than 1 rostral width or 2 annules posterior to rostrum. Buccal cavity with dorsal tooth. Annules on esophageal region without spine-like projections. Six rows of SS on esophageal region, 4 sublateral, 1 dorsal and 1 ventral; 4 sublateral rows on mid-body and annulated tail region. SlAT typical of genus. Anal flap short, crenate. One subdorsal pair of setae about 75% on non-annulated tail region.

Third-Stage Larvae.—L = 0.4 to 0.5 mm. Similar to adults. Rostral setae absent. Amphids doubled spiral, about $3/4$ spiral doubled. CAT typical of genus, posterior tube less than 1 rostral width or 3 annules posterior to rostrum. Eight rows of SS on esophageal region, 4 sublateral, 2 subdorsal and 2 subventral. Five rows of SS on mid-body region, 4 sublateral and 1 dorsal; 4 sublateral rows on annulated tail region. SlAT typical of genus. Anal flap short, crenate. Two to 4 pairs of setae on non-annulated tail region; 1 to 3 subdorsal pairs, 1 long pair 33% to 50% about equal in length to long setae in subdorsal rows opposite PAT, 1 short pair (present or absent) just posterior to last tail annule, 1 short pair (present or absent) about 66%; 1 short lateral to subdorsal pair, usually lateral, adjacent to long subdorsal pair.

Fourth-Stage Larvae.—Similar to adults. Rostral setae present. Amphids doubled spiral, $3/4$ spiral doubled. Eight CAT in 2 transverse rows posterior to rostrum, posterior row less than 1 rostral width or 3 to 5 annules posterior to rostrum. SS not in distinct rows on esophageal region, probably 7, 4 sublateral, 2 subdorsal and 1 ventral. Seven rows of SS on mid-body region, 4 sublateral, 2 subdorsal and 1 ventral; 4 sublateral rows on annulated tail region. Five SlAT and 9 VAT. Anal flap short, crenate. Four to 5 pairs of setae on non-annulated tail region; 1 short lateral to subdorsal pair, usually lateral, about 66%; 2 to 3 subdorsal pairs, 1 short pair (present or absent) just posterior to last complete tail annule, 1 long pair about 33% and equal in length to long setae on annulated tail region, 1 short pair just posterior to second pair; 1 short ventro-sublateral pair adjacent to third subdorsal pair.

Holotype (♂).—Collected February 20, 1965 by B. E. Hopper. Catalogue No. UCNC 1433, University of California, Davis.

Paratypes.—13 ♀♀, 3 ♂♂. Same data as holotype. Deposited at University of California, Davis (UCNC 1519 to 1520); USDA Nematode Collection, Beltsville, Maryland; U. S. National Museum of Natural History, Smithsonian Institution, Washington, D. C.; Station Marine D'Endoune et Centre D'Océanographie, Marseille, France; and Laboratoria voor Morfologie en Systematiek, Museum voor Dierkunde, Gent, Belgium.

Larval Stages.—1 second, 3 third, 9 fourth. Same data as holotype. Deposited at University of California, Davis; USDA Nematode Collection, Beltsville, Maryland; U. S. National Museum of Natural History, Smithsonian Institution, Washington, D. C.; and Station Marine D'Endoune et Centre D'Océanographie, Marseille, France.

Type Habitat.—Marine, associated with *Halimeda* sp., a calcareous alga.

Type Locality.—Coral Key, Florida, USA.

Distribution.—Coral Key and Soldier Key, Florida, USA.

Diagnosis.—Differs from other known species of *Prochaetosoma* by 14 CAT.

Larval stages differ from other known larval stages of *Prochaetosoma* by doubled spiral amphids, second-stage $1/2$ spiral doubled, third- and fourth-stage $3/4$ spiral doubled.

Prochaetosoma vitielloi[28] n. sp.

(Figs. 151 to 155)

Measurements (23 ♀♀): L = 0.6 (0.5–0.7) mm; b = 5.7 (4.9–6.5); c = 11.0 (8.9–13.5); V = 50 (45–54)%; CAT = 24 (21–28) μm; No CAT = 8; SS = 7–91 μm; first SlAT = 45 (41–48) μm; last SlAT = 37 (30–42) μm; first SvAT = 30 (28–33) μm; last SvAT = 27 (23–31) μm; first VAT = 26 (23–29) μm; last VAT = 29 (22–31) μm; No SlAT = 7 (5–8); No SvAT = 3; No VAT = 4 (3–5); Non-ann Term to Tail Length = 61 (55–75)%; T/ABD = 2.6 (2.3–2.9).

(23 ♂♂): L = 0.6 (0.5–0.7) mm; a = 15.5 (12.9–18.2); b = 5.9 (4.9–6.7); c = 10.0 (8.0–13.0); CAT = 25 (22–28) μm; No CAT = 8; SS = 7–99 μm; first SlAT = 43 (35–47) μm; last SlAT = 38 (32–42) μm; first SvAT = 31 (27–33) μm; last SvAT = 28 (23–31) μm; first VAT = 27 (24–31) μm; last VAT = 30 (28–33) μm; No SlAT = 5; No SvAT = 3 (2–3); No VAT = 3 (3–4); Non-ann Term to Tail Length = 53 (37–59)%; T/ABD = 2.8 (2.5–3.2); Spic = 51 (47–56) μm; Gub = 13 (11–16) μm.

(Holotype ♂): L = 0.6 mm; a = 15.5; b = 6.2; c = 8.9; CAT = 26 μm; No CAT = 8; SS = 9–78 μm; first SlAT = 43 μm; last SlAT = 40 μm; first SvAT = 30 μm; last SvAT = 26 μm; first VAT = 26 μm; last VAT = 30 μm; No SlAT = 5; No SvAT = 3; No VAT = 3; Non-ann Term to Tail Length = 52%; T/ABD = 2.9; Spic = 50 μm; Gub = 11 μm.

(4 Fourth-Stage Larvae): L = 0.4–0.5 mm; b = 4.8–6.8; c = 7.0–12.0; No CAT = 4; SlAT = 32–42 μm; VAT = 22–30 μm; No SlAT = 5; No VAT = 3; Non-ann Term to Tail Length = 57–61%; T/ABD = 2.8–2.9.

Male Holotype.—(Figures in parentheses refer to range within species.) Rostrum with faint subcuticular granular markings. Amphids modified loop appearing circular (fig. 153). Eight CAT, paired, in 1 (1 to 2) transverse rows posterior to rostrum (fig. 152), posterior row 1 rostral width or 7 annules (less than 1 rostral width or 6 to 7 annules) posterior to rostrum. Buccal cavity with dorsal tooth, and 1 obscure minute subventral tooth almost adjacent to dorsal tooth. Annules in esophageal region with faint subcuticular granular markings, slightly larger than remaining annules, anterior margins directed anteriorly. Annules with 2 transverse rows of obscure dot-like punctations, most prominent near posterior end of esophagus, few and scattered on remainder of body. Longest SS on esophageal region. Eight rows of SS on esophageal region, 4 sublateral, 2 subdorsal and 2 subventral. Seven rows of SS on mid-body, 4 sublateral, 2 subdorsal and 1 ventral; 4 sublateral rows on annulated tail region. PAT with 4 anterior rows, 2 sublateral and 2 subventral; and 3 posterior rows, 2 sublateral and 1 ventral (fig. 154). Long and short setae intermingled with SlAT. Four long setae with SlAT, not alternating with tubes, setae shorter than adjacent tubes. Caudal glands extend anterior to anus 2.0 (2.0–3.6) times ABD; glands appear abnormal, spinneret absent (fig. 155). Three pairs of anal setae, 19 μm (15 to 20 μm) long, 1 pair anterior and 2 posterior to anus. Anal flap absent (some specimens with short, crenate anal flap). Three (3 to 6) pairs of setae on non-annulated tail region, position measured from last complete tail annule to tail tip; 1 (1 to 2) subventral pairs, 1 long pair about 50% and equal in length to anal setae (some specimens with 1 short pair just posterior to last complete tail annule); 2 (2 to 3) subdorsal pairs, 1 short pair about 13%, 1 short pair about 25%, (some specimens with 1 long pair just posterior to last complete tail annule); (some specimens with 1 short lateral to subdorsal pair, usually lateral, about 50%).

Females.—Similar to males. Posterior row of CAT less than 1 rostral width or 7 to 9 annules posterior to rostrum. Two subventral pairs of paravulval setae, 16 to 21 μm long, 1 anterior and 1 pair posterior to vulva. Longest SS on esophageal region. Short setae intermingled with SlAT. Caudal glands extend anterior to anus 3.4 to 4.6 times ABD. Anal flap short, crenate. Four to 6 pairs of setae on non-annulated tail region; 1 short lateral to subdorsal pair, usually lateral, about

---

28. Named in honor of P. Vitiello.

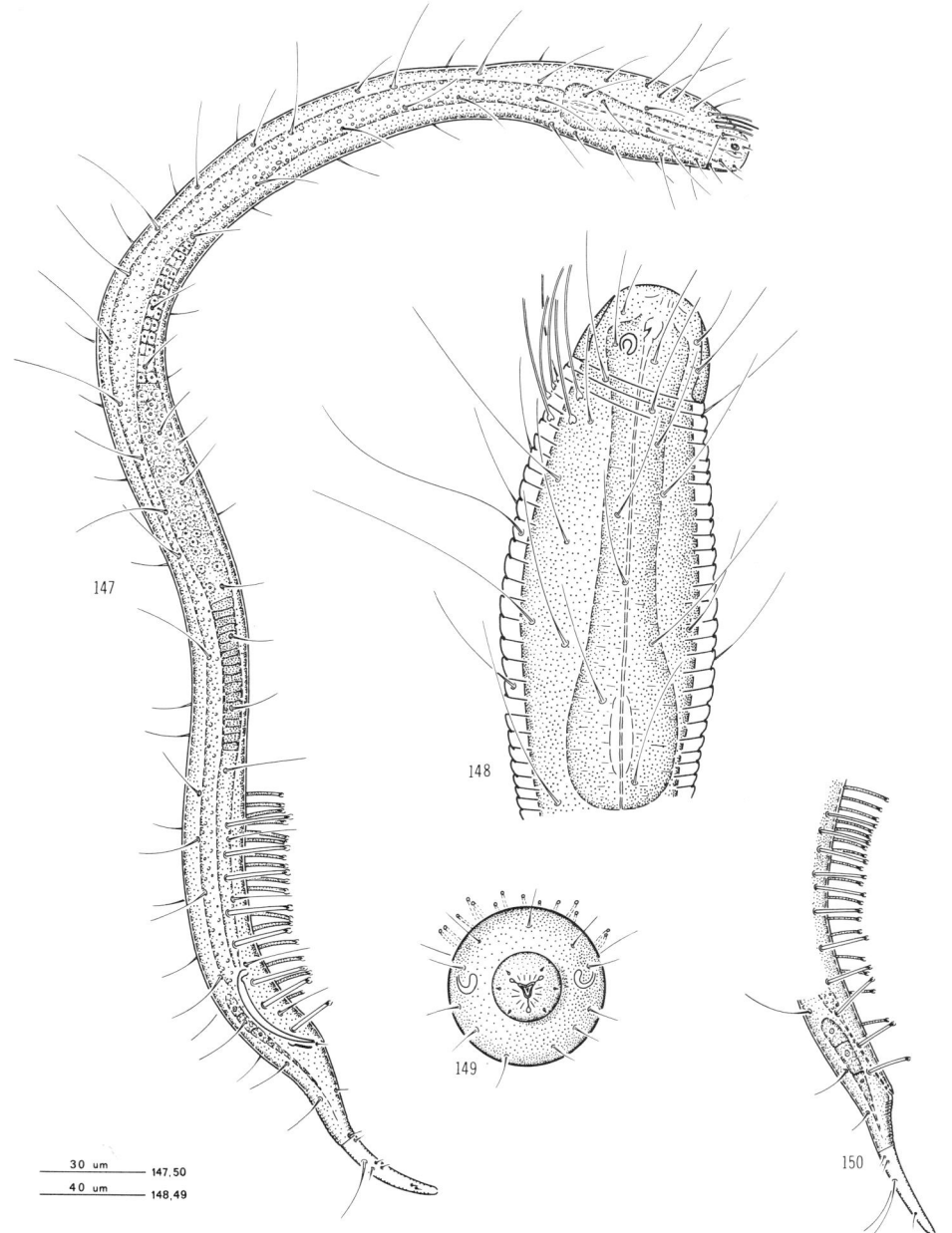

23. Figs. 147-150: *Prochaetosoma cayense* n. sp. 147) Male, full length[a], correct number of CAT not illustrated; 148) Male, head and esophageal region, correct number of CAT not illustrated; 149) Male, face view, subventral tooth not illustrated; 150) Female, PAT[a] and tail.

a. Short SS with SlAT not illustrated.

50%; 3 subdorsal pairs, 1 long pair just posterior to last complete tail annule about equal in length to long setae in subdorsal row opposite PAT, 1 short pair just posterior to first pair or about 13%, 1 short pair about 25%; subventral pairs present or absent, 1 short pair just posterior to last complete tail annule, 1 short pair about 50%.

Second-Stage Larvae.—L = 0.3 mm. Similar to adults. Rostrum ornamentation very obscure, setae absent. CAT typical of genus, tube 1 rostral width or less or 8 to 10 annules posterior to rostrum. Annulation without ornamentation. Six rows of SS on esophageal region, 4 sublateral, 1 dorsal and 1 ventral; 4 sublateral rows on mid-body region. One dorso-sublateral pair of setae on annulated tail region about $1/2$ between anus and last tail annule. SlAT typical of genus. Anal flap short, crenate. Setae on non-annulated tail region present or absent, usually 1 subdorsal pair about 50%, sometimes 1 subventral pair about 33%.

Third-Stage Larvae.—L = 0.3 to 0.4 mm. Similar to adults. CAT typical of genus, posterior tube less than 1 rostral width or 7 to 10 annules posterior to rostrum. Annulation with obscure subcuticular markings. Six rows of SS on esophageal region, 4 sublateral, 1 dorsal and usually only 1 long seta in ventral row. Five rows of SS on mid-body region, 4 sublateral and 1 dorsal; 4 sublateral rows on annulated tail region. SlAT typical of genus. Anal flap crenated. Three pairs of setae on non-annulated tail region; 1 long subdorsal pair just posterior to last complete tail annule about equal in length to long setae in subdorsal row opposite PAT; 1 short lateral to subdorsal pair, usually lateral, about 50%; 1 short subventral pair about 50%.

Fourth-Stage Larvae.—Similar to adults. Four CAT in 1 transverse row posterior to rostrum, row less than 1 rostral width or 8 to 10 annules posterior to rostrum. Annulation with obscure subcuticular markings. Eight rows of SS on esophageal region, 4 sublateral, 2 subdorsal and 2 subventral. Seven rows of SS on mid-body region, 4 sublateral, 2 subdorsal and 1 ventral; 4 sublateral rows on annulated tail region. Five SlAT and 3 VAT. Anal flap short, crenate. Three to 4 pairs of setae on non-annulated tail region; 1 long subdorsal pair just posterior to last complete tail annule about equal in length to long setae in subdorsal row opposite PAT; 2 short lateral to subdorsal pairs, usually subdorsal, 1 pair just posterior to long subdorsal pair, 1 pair 50% to 66%; usually 1 short subventral pair about 50%.

Holotype (♂).—Collected February, 1972 by P. Vitiello. Catalogue No. UCNC 1434, University of California, Davis.

Paratypes.—59 ♀♀, 43 ♂♂. Same data as holotype. Deposited at University of California, Davis (UCNC 1523 to 1525); USDA Nematode Collection, Beltsville, Maryland; U. S. National Museum of Natural History, Smithsonian Institution, Washington, D. C.; Station Marine D'Endoune et Centre D'Océanographie, Marseille, France; and Laboratoria voor Morfologie en Systematiek, Museum voor Dierkunde, Gent, Belgium.

Larval Stages.—3 second, 18 third, 29 fourth. Same data as holotype. Deposited in same nematode collections as paratypes.

Type Habitat.—Marine, collected from medium coarse sand at 18 meters.

Type Locality.—Calanque de Port Miou, near Marseille, France.

Distribution.—Region of Marseille, France.

Diagnosis.—Differs from other known *Prochaetosoma* by PAT in 4 anterior and 3 posterior rows.

Second-stage larvae most closely resemble *P. mediterranicum* n. sp. differ by 1 pair of setae on non-annulated tail region, and 4 sublateral rows of SS on mid-body region. Differ from *P. cayense* n. sp. and *P. arcticum* n. comb. by amphids modified loop appearing circular. Third-stage most closely resemble *P. mediterranicum* n. sp. differ by 1 short lateral pair of setae about 50% and 1 subventral pair on non-annulated tail region. Differ from third-stage *P. cayense* n. sp. by amphids modified loop appearing circular. Fourth-stage larvae differ from other known fourth-stage *Prochaetosoma* by 3 VAT.

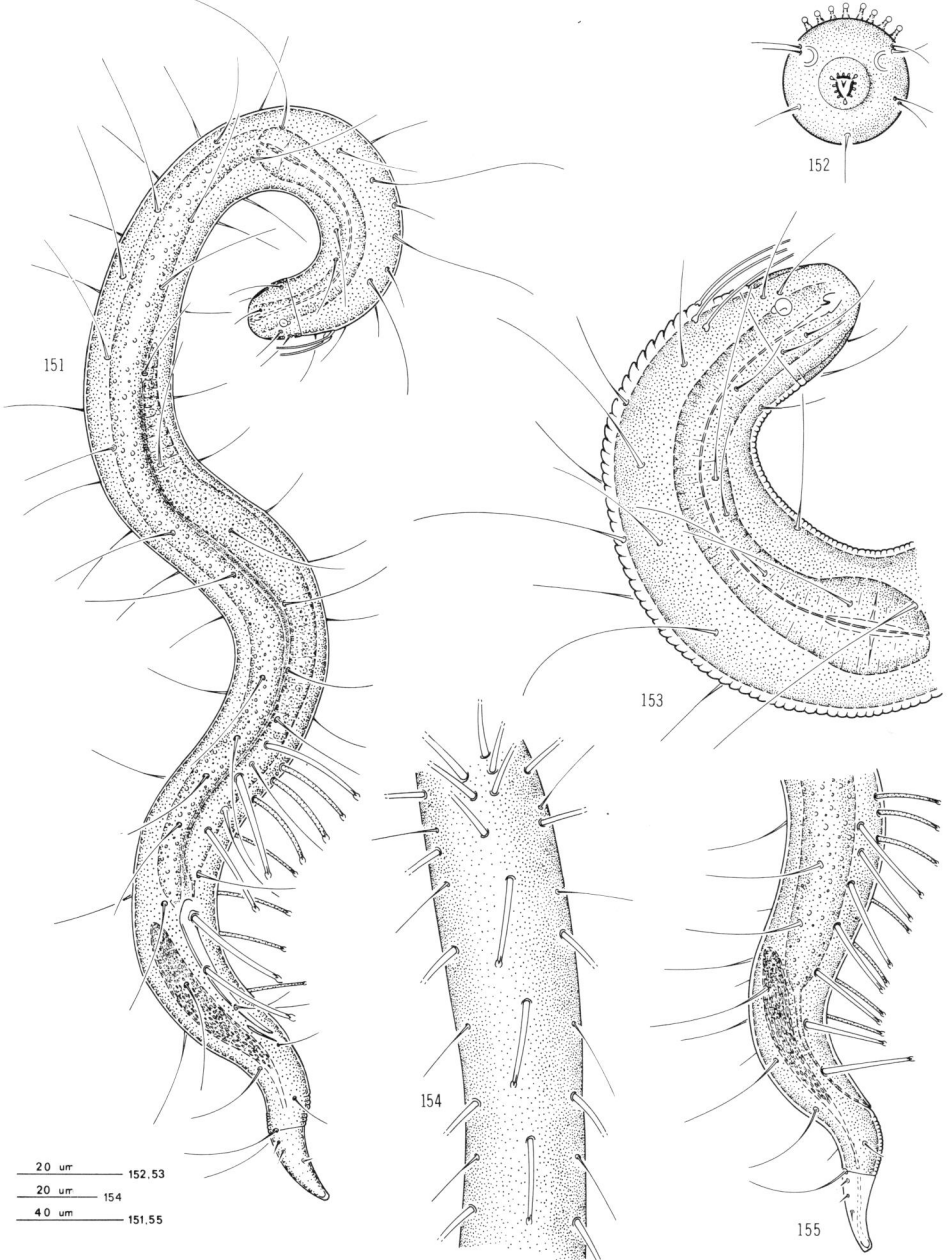

24. Figs. 151-155: *Prochaetosoma vitielloi* n. sp. 151) Male, full length[a], correct number of CAT not illustrated; 152) Male, face view, subventral tooth not illustrated; 153) Male, head and esophageal region, correct number of CAT not illustrated; 154) Male, ventral view of PAT[a], note 4 anterior and 3 posterior rows; 155) Female, posterior body region[a].

a. Short SS with SlAT not illustrated.

Prochaetosoma mediterranicum n. sp.
(Figs. 156 to 159)

Measurements (4 ♀♀): L = 0.7 (0.6–0.8) mm; b = 5.5 (4.8–6.3); c = 9.3 (7.3–10.7); V = 50 (48–52)%; CAT = 26 (26–27) μm; No CAT = 6; SS = 9–90 μm; first SlAT = 50 (46–54) μm; last SlAT = 48 (45–51) μm; first SvAT = 34 (32–37) μm; last SvAT = 32 (29–35) μm; No SlAT = 8 (8–9); No SvAT = 2 (2–3); Non-ann Term to Tail Length = 64 (61–68)%; T/ABD = 2.8 (2.7–3.2).

(12 ♂♂): L = 0.8 (0.6–0.9) mm; a = 15.9 (13.8–18.5); b = 6.3 (5.3–7.3); c = 9.1 (7.8–11.0); CAT = 27 (22–29) μm; No CAT = 6; SS = 8–95 μm; first SlAT = 51 (46–58) μm; last SlAT = 49 (45–51) μm; first SvAT = 36 (32–40) μm; last SvAT = 33 (28–37) μm; No SlAT = 7 (6–8); No SvAT = 2 (2–3); Non-ann Term to Tail Length = 51 (46–56)%; T/ABD = 2.7 (2.1–3.2); Spic = 65 (62–68) μm; Gub = 20 (18–22) μm.

(Holotype ♂): L = 0.6 mm; a = 14.8; b = 6.2; c = 7.8; CAT = 26 μm; No CAT = 6; SS = 9–92 μm; first SlAT = 48 μm; last SlAT = 51 μm; first SvAT = 32 μm; last SvAT = 31 μm; No SlAT = 8; No SvAT = 2; Non-ann Term to Tail Length = 55%; T/ABD = 2.8; Spic = 63 μm; Gub = 21 μm.

(4 Fourth-Stage Larvae): L = 0.5–0.6 mm; b = 4.8–7.0; c = 6.5–10.3; No CAT = 4; SlAT = 33–45 μm; VAT = 24–31 μm; No SlAT = 5; No VAT = 2; Non-ann Term to Tail Length = 59–67%; T/ABD = 2.5–3.1.

Male Holotype.—(Figures in parentheses refer to range within species.) Rostrum with subcuticular granular markings and punctations. Amphids elongate loop-shaped. Six CAT, in 2 transverse rows posterior to rostrum (fig. 158); posterior row 1 rostral width or 11 annules (1 or less rostral widths or 9 to 10 annules) posterior to rostrum. Buccal cavity with dorsal tooth, and 1 distinct subventral tooth (fig. 158). Annules on esophageal region with obscure subcuticular granular markings, slightly larger than annules on mid-body region, margins directed anteriorly. Annules subcuticular granulation obscure on mid-body region; tail annules larger, without ornamentation. Longest SS on esophageal region. Eight rows of SS on esophageal region, 4 sublateral, 2 subdorsal and 2 subventral. Seven rows of SS on mid-body region, 4 sublateral, 2 subdorsal and 1 ventral; 4 sublateral rows on annulated tail region. Four longitudinal rows of PAT, 2 sublateral and 2 subventral. Long and short setae intermingled with SlAT. Three (3 to 4) long setae with SlAT, not alternating with tubes, setae shorter than adjacent tubes. Caudal glands extend anterior to anus 1.4 (1.4 to 2.2) times ABD. Glands appear abnormal, spinneret absent (fig. 159). Three pairs of stout, broad-based anal setae, 11 to 14 μm (10 to 14 μm) long; 1 pair anterior, 1 adjacent to and 1 posterior to anus (some with middle pair slightly posterior to anus). Anal flap short, crenate. Two (2 to 3) pairs of setae on non-annulated tail region, position measured from last complete tail annule to tail tip; 1 long subventral pair about 50% and about equal in length to anal setae; 1 short lateral pair (lateral to subdorsal) about 50% (50% to 75%); (some with 1 short subdorsal pair just posterior to last complete tail annule).

Females.—Similar to males. Amphids modified loop appearing circular (fig. 157). CAT usually in 1 transverse row posterior to rostrum, row 1 or less rostral widths or 8 to 9 annules posterior to rostrum. Two subventral pairs of paravulval setae, 16 to 20 μm long, 1 pair anterior and 1 posterior to vulva. Longest SS on esophageal region. Short setae intermingled with SlAT. Caudal glands extend anterior to anus 1.4 to 2.7 times ABD. Anal flap short, crenate. Two to 3 pairs of setae on non-annulated tail region; 1 subdorsal to lateral pair, usually subdorsal, just posterior to or 3 annule widths posterior to last complete tail annule; 1 lateral pair about 66%; usually 1 subventral to lateral pair, usually subventral, about 66%.

Second-Stage Larvae.—L = 0.3 mm. Similar to adults. Rostrum ornamentation obscure, setae absent. Amphids modified loop, appearing circular. CAT typical of genus, tube less than 1 rostral width or 8 annules posterior to rostrum. Annulation without ornamentation. Six rows of SS on esophageal region, 4 sublateral, 1 dorsal and 1 ventral; SS on remainder of body to anus in 2 dorso-sublateral rows; 1 pair of ventro-sublateral setae adjacent to anus. Usually 2 dorso-sublateral rows of few setae on annulated tail region. SlAT typical of genus. Anal flap short, crenate. Non-annulated tail region usually with single dorsal seta about 50% to 66%.

Third-Stage Larvae.—L = 0.3 to 0.4 mm. Similar to adults. Amphids modified loop, appearing circular. CAT typical of genus, posterior tube less than 1 rostral width or 8 to 9 annules posterior to rostrum. Annulation with subcuticular granulation on esophageal region. Six rows of SS on esophageal region, 4 sublateral, 1 dorsal, usually only 1 long seta in ventral row. Five rows of SS on mid-body region, 4 sublateral, and 1 dorsal; 4 sublateral rows on annulated tail region. SlAT typical of genus. Anal flap short, crenate. Two to 3 pairs of setae on non-annulated tail region; 1 short subventral pair, usually absent, about 50%; 2 subdorsal pairs, 1 long pair just posterior to last complete tail annule; about equal in length to long setae in subdorsal row opposite PAT, and 1 short pair about 50%.

Fourth-Stage Larvae. —Similar to adults. Amphids modified loop, appearing circular. Four CAT, in 2 transverse rows posterior to rostrum, posterior row 1 or less rostral widths or 10 to 11 annules posterior to rostrum. Eight rows of SS on esophageal region, 4 sublateral, 2 subdorsal and 2 subventral. Seven rows of SS on mid-body region, 4 sublateral, 2 subdorsal and 1 ventral; 4 sublateral rows on annulated tail region. Five SlAT and 2 VAT. Anal flap short, crenate. One to 3 pairs of setae on non-annulated tail region; usually 1 short subventral pair 50% to 66%; 1 to 2 subdorsal pairs, 1 long pair (present or absent) just posterior to last tail annule about equal in length to long setae in subdorsal rows opposite PAT, and 1 short pair 50% to 66% (usually slightly posterior to subventral pair).

Holotype (δ).—Collected February, 1972 by P. Vitiello. Catalogue No. UCNC 1435, University of California, Davis.

Paratypes. —2 ♀♀, 10 ♂♂. Same data as holotype. Deposited at University of California, Davis (UCNC 1528 to 1529); USDA Nematode Collection, Beltsville, Maryland; U. S. National Museum of Natural History, Smithsonian Institution, Washington, D. C.; Station Marine D'Endoune et Centre D'Océanographie, Marseille, France; and Laboratoria voor Morfologie en Systematiek, Museum voor Dierkunde, Gent, Belgium.

Larval Stages. —9 second, 4 third, 1 fourth. Same data as holotype. Deposited at University of California, Davis.

Type Habitat. —Marine, collected from medium coarse sand at 18 meters.

Type Locality. —Calanque de Port Miou, near Marseille, France.

Distribution. —Region of Marseille, France.

Diagnosis. —Most closely resembles *P. campbelli* n. comb. differs by fewer SvAT. Differs from *P. vitielloi* n. sp. by PAT in 4 longitudinal rows; from *P. cayense* n. sp. by 6 CAT; and from *P. arcticum* n. comb., *P. longicapitata* n. comb., and *P. lugubre* n. comb. by fewer SlAT and SvAT.

Second-stage larvae most closely resemble *P. vitielloi* n. sp. differ by 2 dorso-sublateral rows of setae on mid-body region, and 1 pair of ventro-sublateral setae adjacent to anus. Differ from *P. cayense* n. sp. and *P. arcticum* n. comb. by amphids a modified loop appearing circular. Third-stage most closely resemble third-stage *P. vitielloi* n. sp. differ by 1 short subdorsal pair of setae about 50% and absence of subventral setae on non-annulated tail region. Differ from *P. cayense* n. sp. by amphids a modified loop appearing circular. Fourth-stage differ from other known species of *Prochaetosoma* by 2 VAT.

Prochaetosoma arcticum (Kreis, 1963) n. comb.

Syn: *Draconema arcticum* Kreis, 1963 n. syn.

Measurements (3 ♀♀): L = 0.7–1.0 mm; b = 6.5–8.4; c = 9.4–11.6; V = 48–49%; CAT = 19–25 μm; No CAT = 8; SS = 12–50 μm; SlAT = 23–44 μm; SvAT = 36–48 μm; No SlAT = 12; No SvAT = 12; Non-ann Term to Tail Length = 17%; T/ABD = 4.2.

Males. —Not observed.

Females. —Rostrum present. Amphids loop-shaped, horseshoe-shaped; dorsal and ventral arms closed together, appearing unispiral. Eight CAT, in 2 transverse rows on rostrum, posterior row less than 1 rostral width posterior to rostrum. Buccal cavity with dorsal tooth. Annules on esophageal region with margins directed anteriorly, remaining body annulation finer. Longest SS on esophageal region. Four longitudinal rows of PAT, 2 sublateral and 2 subventral. Caudal glands, spinneret, and non-annulated tail region present.

Second-Stage Larva. —L = 0.4 mm. Similar to adults. Rostrum present, setae absent. Amphids circular loop-shaped. CAT typical of genus, tube 1 rostral width posterior to rostrum. Six rows of SS on esophageal region, 4 sublateral, 1 dorsal and 1 ventral; 4 sublateral rows on mid-body region.

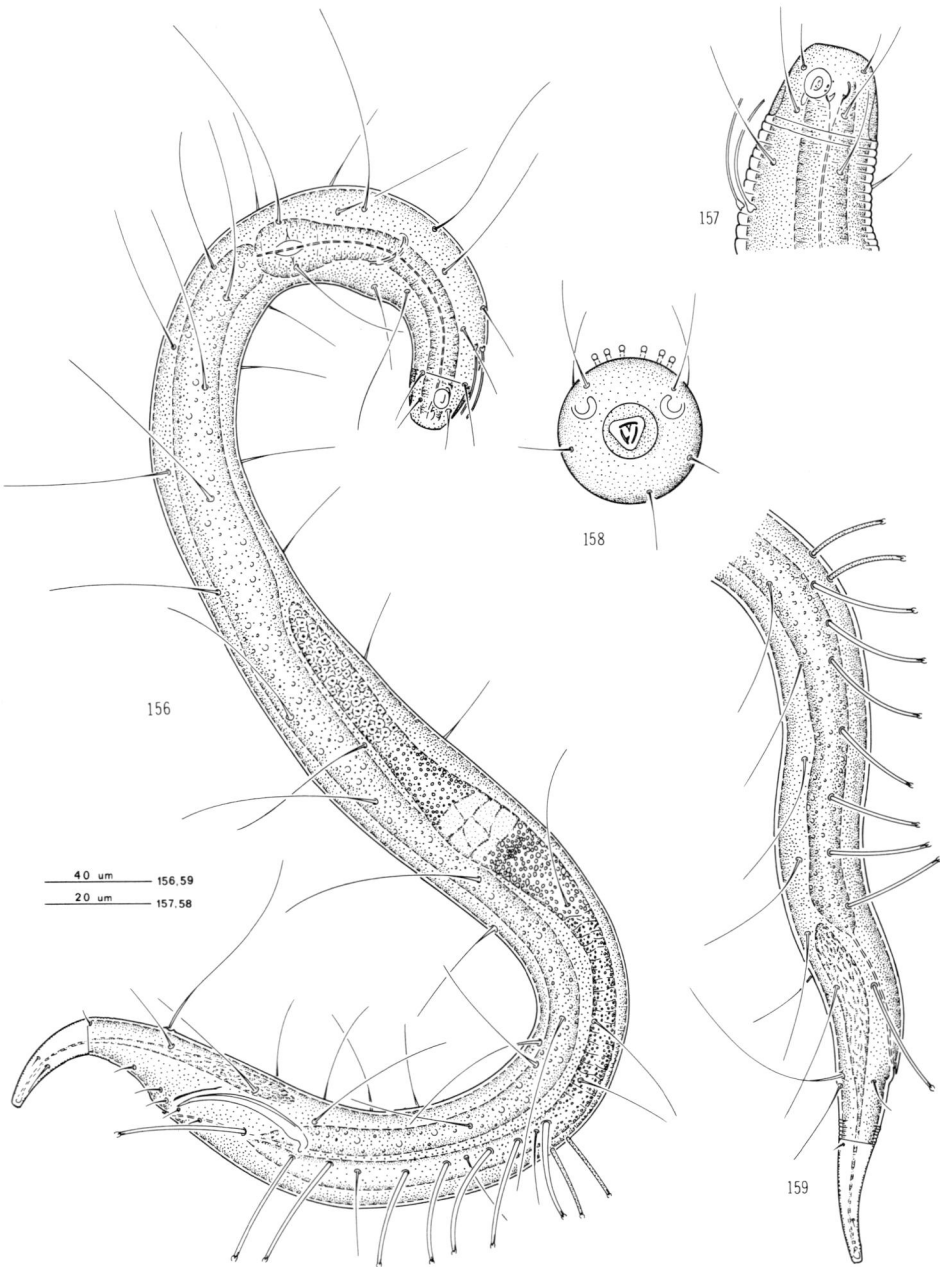

25. Figs. 156–159: *Prochaetosoma mediterranicum* n. sp. 156) Male, full length[a], correct number of CAT not illustrated; 157) Female, head, correct number of CAT not illustrated; 158) Male, face view; 159) Female, posterior body region[a].

a. Short SS with SlAT not illustrated.

One pair of dorso-sublateral setae, on annulated tail region, about $\frac{1}{2}$ ABD posterior to anus. SlAT typical of genus.

Third-Stage Larvae. – Not observed.
Fourth-Stage Larvae. – See discussion following diagnosis.
Type Habitat. – Marine.
Type Locality. – Along banks of Eyjafjörður Fjord, starting at harbor of Akureyri, Iceland.
Distribution. – Same as type locality.
Diagnosis. – Most closely resembles *P. lugubre* n. comb. females differ by absence of stout ventral setae anterior to SvAT, and 12 SvAT. Differs from *P. cayense* n. sp. by 8 CAT; from *P. vitielloi* n. sp. by PAT in 4 longitudinal rows; and from *P. mediterranicum* n. sp. and *P. campbelli* n. comb. by 12 SlAT.

Second-stage larvae differ from other known second-stage larvae of *Prochaetosoma* by elongate loop-shaped amphid.

Type material of this species was not available for study. From the original descriptions of the females and second-stage larva, we were able to distinguish this species from other known species of *Prochaetosoma*.

Discussion. – In 1963, H. A. Kreis described *Draconema arcticum* from Eyjaförður Fjord, Iceland; using 3 females, 1 preadult, and 1 second-stage larva. The females and second-stage larva were illustrated and adequately described.

A premature (preadult) stage was described from a single specimen with 4 longitudinal rows of PAT, neither a vulva nor spicules were indicated, and no illustrations were given for this specimen. Other diagnostic characters given for this 'preadult' were as follows: No CAT = 8; No SlAT = 11; and No SvAT = 6. Taxonomic characters given in the original description for this 'preadult' did not enable us to make a definite species identification for the following reasons. In the numerous specimens of fourth-stage larvae (preadult) studied with normal adhesion tubes, all fourth-stage larvae examined had only 3 longitudinal rows of PAT, never 4 rows. Adults examined in this superfamily always possessed a greater number of CAT and PAT than the fourth-stage larvae. Even if the single row of VAT were mistaken for the 2 rows of SvAT, it is unlikely this 'preadult' would belong to this species. But, for convenience we are retaining it as *P. arcticum* until more specimens can be obtained of this 'preadult' form.

Prochaetosoma campbelli (Allgén, 1932) n. comb.

Syn: *Chaetosoma campbelli* Allgén, 1932 n. syn.
Drepanonema campbelli (Allgén, 1932) Cobb, 1933 n. syn.
Claparediella campbelli (Allgén, 1932) Filipjev, 1934 n. syn.
Draconema campbelli (Allgén, 1932) Schuurmans Stekhoven, 1935 n. syn.

Measurements (3 ♀♀): L = 0.8 mm; b = 7.7 (7.1-8.2); c = 11.1 (10.3-11.6); V = 50%?; CAT = ?; No CAT = 4; SS = ?; SlAT = ?; SvAT = ?; No SlAT = 9; No SvAT = 10-11; Non-ann Term to Tail Length = ?; T/ABD = ?.

(1 ♂): L = 0.8 mm; a = 13.9; b = 8.1; c = 8.4; CAT = 21 μm; No CAT = 4; SS = 30 μm; first SlAT = 33 μm; last SlAT = 29 μm; SvAT = 21 μm; No SlAT = 5; No SvAT = 7-8; Non-ann Term to Tail Length = 50%; T/ABD = 2.8; Spic = 71 μm; Gub = 29 μm.

(2 Fourth-Stage Larvae): L = 0.6 mm; b = 7.6-7.7; c = 9.7-10.2; No CAT = ?; SlAT = ?; VAT = ?; No SlAT = 5; No VAT = 9; Non-ann Term to Tail Length = ?; T/ABD = ?.

Male. – Rostrum present. Amphids circular unispiral. Four CAT, in 2 transverse rows, anterior row on rostrum, posterior row less than 1 rostral width posterior to rostrum. Buccal cavity with dorsal tooth. Annules slightly larger on esophageal region than remaining body annules. SS in distinct rows. Longest SS on annulated tail region. Four longitudinal rows of PAT, 2 sublateral and 2 subventral. Two ventral, stout, conically shaped, broad-based setae just posterior to last SvAT, distal ends directed posteriorly, setae 15 to 20 μm long, width of bases about $\frac{1}{4}$ to $\frac{1}{3}$ setae length. One pair of anal setae, 14 μm long, posterior to anus. Caudal glands, spinneret, and non-annulated tail region present.

Females.—Similar to males. Two subventral pairs of paravulval setae, 10 μm long, 1 pair anterior and 1 posterior to vulva.

Second- and Third-Stage Larvae.—Not observed.

Fourth-Stage Larvae.—Similar to adults. Amphids circular unispiral. SS present. Five SlAT, and 9 VAT.

Type Habitat.—Marine, associated with Red Algae at 40 meters.

Type Locality.—Persev. Harbor, Campbell Island (New Zealand).

Distribution.—Same as type locality.

This species has also been reported from Norway.

Diagnosis.—Differs from other known species of *Prochaetosoma* by 4 CAT; and 2 ventral, stout, conically shaped, broad-based setae just posterior to SvAT.

Fourth-stage larvae most closely resemble fourth-stage *P. cayense* n. sp. differ by unispiral amphids. Differ from *P. primitiva, P. vitielloi* n. sp., and *P. mediterranicum* n. sp. by 9 VAT.

Type material of this species was not available for study. From the original description, we were able to distinguish this species from other known species of *Prochaetosoma*. In the original description it was noted that this species was similar to *P. primitiva* (= *Chaetosoma primitivum* Steiner, 1916).

Discussion.—In the original description of the adults it was stated that the PAT are arranged in 3 longitudinal rows, 2 sublateral and 1 ventral. Because of the great number of specimens examined in this study, it is evident that the 2 rows of SvAT were mistaken as being a single row of adhesion tubes. The 2 ventral, stout, conically shaped, broad-based setae just posterior to last SvAT are probably similar to corniform adhesion tubes (see Morphology, p. 13).

The fourth-stage larvae as described, were lacking CAT. In the numerous specimens of fourth-stage larvae we have examined, there are at least 4 CAT present, total number depends on the number of tubes present on adult specimens. Therefore, we conclude that the CAT were probably broken off or overlooked.

Prochaetosoma longicapitata (Allgén, 1932) n. comb.

Syn: *Chaetosoma longicapitata* Allgén, 1932 n. syn.

Drepanonema longicapitata (Allgén, 1932) Cobb, 1933 n. syn.

Claparediella longicapitata (Allgén, 1932) Filipjev, 1934 n. syn.

Draconema longicapitata (Allgén, 1932) Schuurmans Stekhoven, 1935 n. syn.

Measurements (1 ♂): L = 0.9 mm; a = 15.8; b = 5.7; c = 7.8; CAT = 30 μm; No CAT = 6; SS = 18-30 μm; first SlAT = 33 μm; last SlAT = 21 μm; first SvAT = 25 μm; last SvAT = 10 μm; No SlAT = 18; No SvAT = 4-5; Non-ann Term to Tail Length = 18%; T/ABD = 5.6; Spic = 45 μm; Gub = 10 μm.

Male.—Rostrum present. Amphids loop-shaped, horseshoe-shaped. Six CAT, in 2 transverse rows posterior to rostrum, posterior row less than 1 rostral width posterior to rostrum. Buccal cavity with dorsal tooth. Annulation without ornamentation. Longest SS on body region. Four longitudinal rows of PAT, 2 sublateral and 2 subventral. Caudal glands, spinneret, and non-annulated tail region present.

Females and Larval Stages.—Not observed.

Type Habitat.—Marine, associated with Red Algae at 40 meters.

Type Locality.—Persev. Harbor, Campbell Island (New Zealand).

Distribution.—Same as type locality.

Diagnosis.—Males differ from other known species of *Prochaetosoma* by greater number of SlAT.

Type specimens of this species were not available for study. From the original description we were able to distinguish *P. longicapitata* from the other known species of *Prochaetosoma*. In the description it was indicated that this species is similar to *P. primitiva* (= *Chaetosoma primitivum* Steiner, 1916).

Prochaetosoma lugubre (Gerlach, 1957) n. comb.

Syn: *Drepanonema lugubre* Gerlach, 1957 n. syn.

Measurements (1 ♀): L = 0.7 mm; b = 5.9; c = 10.8; V = 46%; CAT = ?; No CAT = ?; SS = 35-50 μm; SlAT = 17 μm; SvAT = ?; No SlAT = 15; No SvAT = 4-5; Non-ann Term to Tail Length = 57%; T/ABD = 2.8.

Males.—Not observed.

Female.—Rostrum present. Amphids unispiral. (For explanation of CAT, see Discussion.) Buccal cavity with 1 dorsal tooth; 2 minute ventral teeth, 1 ventral and 1 subventral. Annulation without ornamentation. SS in distinct rows. Four longitudinal rows of PAT, 2 sublateral and 2 subventral. Compared to SlAT size; SvAT small, obscure, almost rudimentary. Five ventral stout setae anterior to SvAT. Caudal glands and spinneret present.

Larval Stages.—Not observed.

Type Habitat.—Marine, collected from medium coarse sand and bottom debris at 1.8 meters.

Type Locality.—Beach in front of City of Ilha Bela, São Sebastião Island, State of São Paulo, Brazil.

Distribution.—Same as type locality.

Diagnosis.—Females differ from other known species of *Prochaetosoma* by greater number of SlAT; and ventral, stout setae just anterior to SvAT.

Type material of this species was not available for study. From the original description of the female, we were able to distinguish this species from other known species of *Prochaetosoma*.

Discussion.—In the description, the absence of CAT was used as a diagnostic taxonomic character. Except for first-stage larvae, CAT have been present on all of the specimens examined in this study, so we conclude that the CAT were broken off. In the original description there was no indication of the 3 teeth in the buccal cavity, 1 dorsal and 2 subventral, but these teeth were clearly indicated in the illustrations.

Genus *Draconactus*[29] n. gen.

Diagnosis: Prochaetosomatinae. Nematodes 0.5 to 0.7 mm long. Most relaxed specimens assume an "S" shape. Esophageal region slightly swollen, dorsal side with prominent swelling (fig. 161). Greatest body width at mid-body in females; at esophageal and PAT region in males. Rostrum conical shaped (fig. 161), without ornamentation, setae present. Amphids conspicuous, loop-shaped, dorso-lateral on rostrum. Eight CAT, posterior to rostrum. Buccal cavity weakly developed, partially collapsed; with conspicuous or inconspicuous dorsal tooth. Posterior esophageal bulb without cuticularized valve. Cuticle not unusually thick. Annules without ornamentation. Long and short SS in distinct rows on body, some rows with alternating long and short setae. Paravulval setae absent. Four longitudinal rows of PAT, 2 sublateral and 2 subventral. Males with uniformly tapered anal setae. Short anal flap present. Tails cylindrical-conoid. Spinneret present. Setae present on non-annulated tail region.

Type Species: *Draconactus cutus* n. sp.

Differs from other known genera in *Prochaetosomatinae* by conical shaped rostrum, and prominent dorsal swelling on esophageal region. Differs from *Prochaetosoma* Micoletzky, 1922 by absence of cuticularized valve in posterior esophageal bulb; from *Apenodraconema* n. gen. by loop-shaped amphids, and annules without ornamentation.

Larval Stages.—No first-, second-, or third-stage larvae available in *Draconactus*. Fourth-stage larvae 0.8 mm long. Similar to adults. Widest part of body in esophageal region. Amphids loop-shaped, dorso-lateral on rostrum. Four CAT, posterior to rostrum. Annulation without ornamentation. SS in distinct rows on body, some rows with alternating long and short setae. Three longitudinal rows of PAT, 2 sublateral and 1 ventral. Non-annulated tail region present. Differentiated from other stages by 4 CAT, and 3 rows of PAT.

---

29. *Draconactus*, L. m., thick dragon.

Draconactus cutus n. sp.

(Figs. 160 to 163)

Measurements (3 ♀♀): L = 0.7 (0.6-0.7) mm; b = 10.1 (9.3-11.5); c = 13.2 (9.3-16.5); V = 50 (47-52)%; CAT = 17 (16-19) μm; SS = 7-23 μm; first SlAT = 18 (17-19) μm; last SlAT = 16 (16-17) μm; first SvAT = 15 (12-17) μm; last SvAT = 10 μm; No SlAT = 12 (11-12); No SvAT = 13 (12-15); Non-ann Term to Tail Length = 59 (59-60)%; T/ABD = 3.1 (2.9-3.3).

(1 ♂): L = 0.5 mm; a = 10.4; b = 8.0; c = 9.6; CAT = 18 μm; SS = 9-16 μm; first SlAT = 17 μm; last SlAT = 16 μm; first SvAT = 17 μm; last SvAT = 12 μm; No SlAT = 6; No SvAT = 14; Non-ann Term to Tail Length = 42%; T/ABD = 2.5; Spic = 54 μm; Gub = 18 μm.

(Holotype ♀): L = 0.7 mm; b = 9.4; c = 16.5; V = 47%; CAT = 17 μm; SS = 8-23 μm; first SlAT = 19 μm; last SlAT = 16 μm; first SvAT = 17 μm; last SvAT = 10 μm; No SlAT = 12; No SvAT = 13; Non-ann Term to Tail Length = 59%; T/ABD = 2.9.

Female Holotype.—(Figures in parentheses refer to range within species.) Rostrum typical of genus. Amphids elongate loop-shaped, crescent-shaped; ends of amphidial arms slender, anterior part of loop swollen (fig. 161). Eight CAT, paired, in 3 transverse rows posterior to rostrum, posterior row 1 rostral width or 7 annules posterior to rostrum. Buccal cavity with small inconspicuous dorsal tooth. Annules larger on esophageal and tail regions than intervening annules. Annules on esophageal region with anterior margins directed anteriorly. Annules posterior to PAT with posterior margins directed posteriorly. Eight (8 to 9) tail annules. Longest SS on esophageal region. Eight rows of SS on esophageal and mid-body regions, 4 sublateral, 2 subdorsal and 2 subventral; 4 sublateral rows on annulated tail region. Short setae intermingled with SlAT. Caudal glands extend anterior to anus 5.2 (3.1-5.2) times ABD. Two pairs of setae on non-annulated tail region, position measured from last complete tail annule to tail tip; 1 lateral pair about 50%; and 1 lateral (lateral to subdorsal, usually subdorsal) pair about 66%.

Male.—Similar to females. CAT posterior row slightly more than 1 rostral width posterior to rostrum. Fourteen tail annules. Longest SS on esophageal region. Long setae intermingled with SlAT, short setae absent, alternating with tubes and longer than adjacent tubes. Two pairs of anal setae, 10 to 12 μm long, both pairs anterior to anus. Caudal glands extend anterior to anus 3.1 times ABD. Three pairs of setae on non-annulated tail region, 1 subventral pair just posterior to last complete tail annule, 1 latero-subventral pair about 1 annule width posterior to subventral pair, and 1 subdorsal pair about 50%.

Larval Stages.—Not observed.

Holotype (♀).—Collected June 3, 1968 by I. M. Newell. Catalogue No. UCNC 1436, University of California, Davis.

Paratype.—1 ♂. Same data as holotype. Deposited at University of California, Davis (UCNC 1533).

Type Habitat.—Marine, associated with Brown Algae and Coral.

Type Locality.—Just behind outer edge of reef near Coco Solo, Galeta Beach, Panama, on Caribbean.

Distribution.—Same as type locality.

Diagnosis.—Differs from other known species of *Draconactus* by small inconspicuous dorsal tooth, and smaller size.

Draconactus suillus (Allgén, 1932) n. comb.

Syn: *Chaetosoma suilla* Allgén, 1932 n. syn.
   Drepanonema suilla (Allgén, 1932) Cobb, 1933 n. syn.
   Claparediella suilla (Allgén, 1932) Filipjev, 1934 n. syn.
   Draconema suilla (Allgén, 1932) Schuurmans Stekhoven, 1935 n. syn.

Measurements (1 Fourth-Stage Larva): L = 0.8 mm; b = 9.0; c = 9.5; No CAT = 4; SlAT = 25-29 μm; VAT = 13-19 μm; No SlAT = 5; No VAT = 8; Non-ann Term to Tail Length = 40%; T/ABD = 3.6.

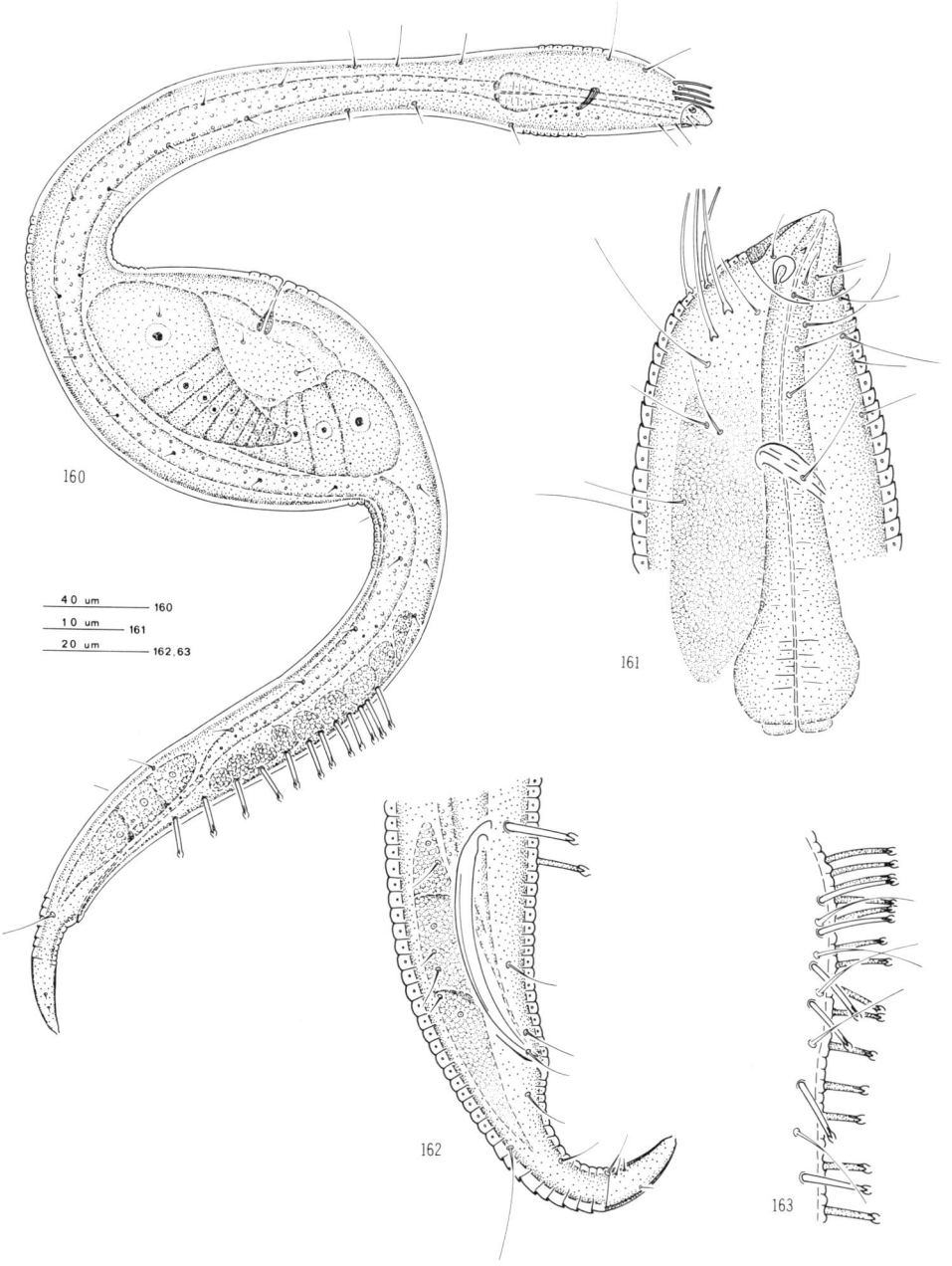

26. Figs. 160-163: *Draconactus cutus* n. sp. 160) Female, full length[a], correct number of SvAT not illustrated; 161) Female, head and esophageal region; 162) Male, tail; 163) Male, PAT.

a. Short SS with SlAT not illustrated.

Males, Females, Second-, and Third-Stage Larvae.—Not observed.

Fourth-Stage Larva.—Rostrum typical of genus. Amphids loop-shaped, horseshoe-shaped. Four CAT, less than 1 rostral width posterior to rostrum. Buccal cavity with large conspicuous dorsal tooth. Annulation without ornamentation. Annules wider on esophageal and tail regions, than intervening annules. Thirteen tail annules. SS in distinct rows on body, longest on esophageal region. Five SlAT and 8 VAT.

Type Habitat.—Marine, associated with Red Algae at 40 meters.
Type Locality.—Persev. Harbor, Campbell Island (New Zealand).
Distribution.—Same as type locality.
Diagnosis.—Differs by large conspicuous dorsal tooth, larger size, and number of VAT.

Type material of this species was not available for study. Because of the stable taxonomic characters in fourth-stage larvae of the superfamily, and distinctive characters given in the original description of this specimen, we conclude that *D. suillus* is a good species.

Discussion.—The number of CAT was not given in the description, but the tubes were clearly indicated on the illustration. None of the fourth-stage larvae examined in this superfamily has been larger than adults. Fourth-stage larva of *D. suillus* is larger than adults of *D. cutus* n. sp., so it can be assumed that adults of this species will also be much larger than *D. cutus* n. sp.

Genus *Apenodraconema*[30] n. gen.

Diagnosis: Prochaetosomatinae. Nematodes 0.7 to 0.9 mm long. Most relaxed specimens assume an open "S" shape. Anterior body region sometimes distinctively arched ventrally. Esophageal region slightly swollen, greatest width at mid-body in females. Rostrum broadly rounded anteriorly, setae present. Rostrum with conspicuous vacuoles, most prominent just anterior to first body annule. Amphids conspicuous, spiral, dorso-lateral on rostrum. Eight CAT, paired, in 2 transverse rows posterior to rostrum. Buccal cavity weakly developed, partially collapsed; with inconspicuous dorsal tooth. Posterior esophageal bulb without cuticularized valve. Cuticle not unusually thick. Margins of annular ridges with obscure or prominent spine-like projections (fig. 166). SS in distinct rows on body, some rows with alternating long and short setae. Females with or without paravulval setae. Four longitudinal rows of PAT, tubes variable in number between species. Anal flap present. Tails elongate cylindrical-conoid, spike-like, non-annulated tail region 90% total tail length; 3 to 4 tail annules (fig. 164). Spinneret present. Setae on non-annulated tail region.

Type Species: *Apenodraconema chlidosis* n. sp.

Differs from other known genera in Prochaetosomatinae by elongate cylindrical-conoid, spike-like tail; 3 to 4 tail annules; and spine-like projections on margins of annular ridges. Differs from *Prochaetosoma* Micoletzky, 1922 by absence of cuticularized valve in posterior esophageal bulb; from *Draconactus* n. gen. by spiral amphid.

Larval Stages: No first-, second-, or third-stage larvae available in *Apenodraconema*. Fourth-stage larvae 0.6 mm long. Similar to adults. Widest part of body in esophageal region. Amphids conspicuous, spiral. Four CAT, in 1 transverse row posterior to rostrum. SS in distinct rows on body. Three longitudinal rows of PAT, 2 sublateral and 1 ventral. Anal flap present.

Key to Species of *Apenodraconema* (Females)

1. Amphids unispiral, V at 44% . . . . *chlidosis* n. sp. (p. 109)
   Amphids double spiral, V at 49%. . . . . *spinicaudum* (Gerlach, 1958) n. comb. (p. 109)

---

30. *Apenodraconema*, Gr. n., rough dragon nematode.

Apenodraconema chlidosis[31] n. sp.
(Figs. 164 to 166)

Measurements (Holotype ♀): L = 0.7 mm; b = 8.8; c = 7.0; V = 44%; CAT = 25 μm; SS = 8-34 μm; first SIAT = 39 μm; last SIAT = 34 μm; first SvAT = 40 μm; last SvAT = 28 μm; No SIAT = 6; No SvAT = 11; Non-ann Term to Tail Length = 90%; T/ABD = 6.9.

Female Holotype.—Anterior body region distinctively arched ventrally. Amphids unispiral. Eight CAT, posterior row less than 1 rostral width or 3 annules posterior to rostrum. Buccal cavity with inconspicuous dorsal tooth. Margins of annular ridges with prominent spine-like projections (fig. 166). Three tail annules. Two subventral pairs of paravulval setae, 10 μm long, 1 pair anterior and 1 posterior to vulva. Longest SS on non-annulated tail region. Eight rows of SS on esophageal and mid-body regions, 4 sublateral, 2 subdorsal and 2 subventral. Setae absent on annulated tail region. Medium length setae alternate with anterior SIAT, about equal in length to longest setae in subdorsal row opposite PAT. Short setae alternating with posterior SIAT. Caudal glands extend anterior to anus 3.1 times ABD. Long anal flap present. Five pairs of setae on non-annulated tail region, position measured from last complete tail annule to tail tip; 1 short latero-subventral pair about 50%; 1 short latero-subdorsal pair about 3 annule widths anterior to tail terminus; 3 subdorsal pairs, 1 long pair about 25% and equal in length to longest setae on esophageal region, 1 short pair just posterior to first pair, and 1 short pair 1 annule width posterior to second pair.

Males, and Larval Stages.—Not observed.

Holotype (♀).—Collected January 20, 1968 by I. M. Newell. Catalogue No. UCNC 1437, University of California, Davis.

Paratypes.—None.

Type Habitat.—Marine, associated with Red Corallines and other Algae.

Type Locality.—Forty-three kilometers east of Papeete, Tahiti Island, Society Islands.

Distribution.—Same as type locality.

Diagnosis.—Females differ from *A. spinicaudum* n. comb. by unispiral amphid and V at 44%.

Apenodraconema spinicaudum (Gerlach, 1958) n. comb.

Syn: *Draconema spinicaudum* Gerlach, 1958 n. syn.

Measurements (Holotype ♀): L = 0.9 mm; b = 8.0; c = 7.1; V = 49%; CAT = 28 μm; SS = 10-52 μm; PAT = 31-40 μm; No SIAT = 5; No SvAT = 11; Non-ann Term to Tail Length = 90%; T/ABD = 6.5.

Female Emended.—Anterior body region not distinctively arched ventrally. Amphids double spiral. Eight CAT, posterior row less than 1 rostral width or 5 annules posterior to rostrum. Buccal cavity with inconspicuous dorsal tooth. Margins of annular ridges with obscure spine-like projections, most prominent on posterior margins. Four tail annules. Paravulval setae absent. Longest SS on non-annulated tail region. Eight rows of SS on esophageal and mid-body regions, 4 sublateral, 2 subdorsal and 2 subventral; 4 sublateral rows on annulated tail region. Short and medium length setae intermingled with SIAT. Anal flap present. Five pairs of setae on non-annulated tail region, position measured from last complete tail annule to tail tip; 1 short subventral pair about 25%; 4 subdorsal pairs, 1 short pair 2 to 3 annule widths posterior to last complete tail annule, 1 long pair (52 μm long) about 33%, 1 short pair 1 annule width posterior to second pair, 1 short pair 1 annule width posterior to third pair.

Males, First-, Second-, and Third-Stage Larvae.—Not observed.

Fourth-Stage Larva Emended.—L = 0.6 mm. Similar to adults. Amphids double spiral. Four CAT, in 1 transverse row less than 1 rostral width or 4 annules posterior to rostrum. Seven tail annules. Eight indistinct rows of SS on esophageal region, 4 sublateral, 2 subventral and 2 subdorsal. Seven rows of SS on mid-body region, 4 sublateral, 2 subdorsal and 1 ventral; 4 sublateral rows on annulated tail region. Five SIAT and 9 VAT. Three pairs of setae on non-annulated tail region; 1

---

31. *chlidosis*, ornamentation.

short subventral pair about 50%; 2 subdorsal pairs, 1 long pair (52 μm long) about 33%, 1 short pair 1 annule width posterior to first pair.

Holotype (♀).—Collected in Sample X13, November 14, 1957. Deposited at Institut für Meeresforschung, Bremerhaven, Germany.

Paratype.—1 fourth-stage larva. Slide No. 180d. Same data as holotype. Deposited in same nematode collection as holotype.

Type Habitat.—Marine, associated with *Tubipora* sp.

Type Locality.—A coral reef, Sarso Island, Red Sea.

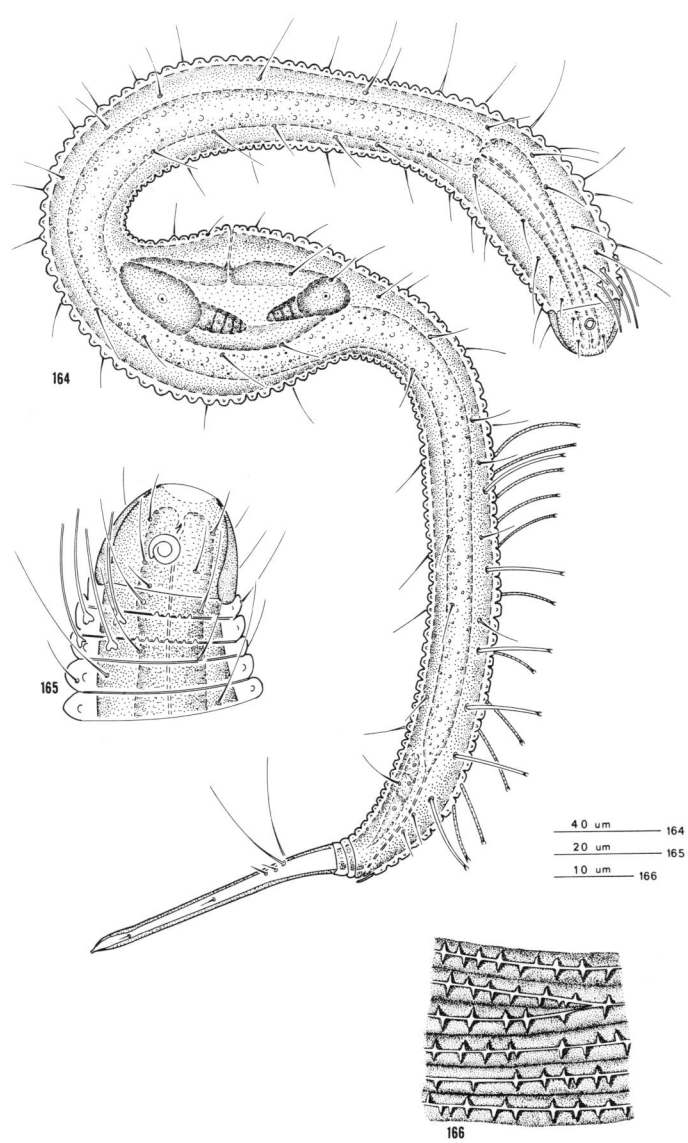

27. Figs. 164–166: *Apenodraconema chlidosis* n. sp. 164) Female, full length[a]; 165) Female, head; 166) Female, cuticle, note annular ridges with spine-like projections.

a. Short SS with SlAT not illustrated.

Distribution.—Same as type locality.

Diagnosis.—Differs from females of *A. chlidosis* n. sp. by double spiral amphids and V at 49%. The paratype specimen (fourth-stage larva) of this species was available for study, and was in good condition. With the aid of the original description and paratype specimen, this species was easily identified.

Discussion.—The inconspicuous dorsal tooth was not indicated in the original description, but was present in the fourth-stage paratype larva. Because of the very stable taxonomic characters of fourth-stage larvae in the superfamily, we conclude that the inconspicuous dorsal tooth is also present in the female.

Genus *Cygnonema*[32] n. gen.

Diagnosis: Cygnonematinae. Nematodes 1.7 to 3.1 mm long. Anterior 50% to 59% of body long and slender; relaxed anterior body usually arched posteriorly, then anteriorly, and then arched posteriorly (fig. 167). Slightly swollen at base of esophagus. Greatest body width in both sexes in posterior 41% to 50% of body. Rostrum broadly rounded anteriorly, with faint markings and subcuticular granulation, setae present. Amphids dorso-lateral on rostrum. Male amphids conspicuous, elongate loop-shaped, dorsal arm elongated (fig. 169). Female amphids inconspicuous, tubular-shaped, a simple elongate tube, without anterior loop of male amphids. CAT usually paired, in 3 or more transverse rows, posterior to rostrum; posterior tube 2 to 3 rostral widths or 26 to 37 annules posterior to rostrum. Buccal cavity weakly developed, with conspicuous dorsal tooth. Esophagus with elongate cylindrical corpus, terminal swelling without cuticularized valve. Cuticle not unusually thick. Annules with obscure subcuticular granulation. SS in distinct rows on body, some rows with alternating long and short setae, setae not intermingled with SlAT. Paravulval setae short, stout, tube-like (fig. 170). Four longitudinal rows of PAT, 2 sublateral and 2 subventral. Males with uniformly tapered anal setae. Anal flap present. Setae on non-annulated tail region.

Type Species: *Cygnonema steineri* n. sp.

Differs by long, slender anterior body region (50% to 59% total body length), and location of CAT (see Subfamily Diagnosis p. 18).

Larval Stages: No first-stage larvae available in *Cygnonema*. Second-stage larvae 0.7 to 1.0 mm long. Similar to adults. Swollen regions on body not as prominent as adults. Rostrum ornamentation obscure. Amphids as in females, inconspicuous, tubular-shaped. Single dorsal CAT, usually posterior to rostrum, sometimes on rostrum. Annulation without ornamentation. SS in distinct rows on body, some rows with alternating long and short setae. Two pairs of SlAT, in 2 longitudinal rows. Anal flap absent. Setae on non-annulated tail region. Differentiated from other 2 stages by 1 CAT, 2 pairs of SlAT in 2 rows.

Third-stage larvae 1.1 to 1.4 mm long. Similar to adults. Swollen body regions not as prominent as adults. Amphids as in females, inconspicuous, tubular-shaped. Three CAT, 2 subdorsals and 1 dorsal, anterior tube usually on rostrum and posterior tube posterior to rostrum. Annulation without ornamentation. SS in distinct rows on body, some rows with alternating long and short setae. Two longitudinal rows of PAT, number of SlAT variable. Anal flap present or absent. Differentiated from other 2 stages by 3 CAT, 2 rows of SlAT.

Fourth-stage larvae 1.7 to 2.2 mm long. Similar to adults. Swollen body regions not as prominent as adults. Amphids as in females, inconspicuous, tubular-shaped. Four to 6 CAT, usually posterior to rostrum. SS in distinct rows on body, some rows with alternating long and short setae. Three longitudinal rows of PAT, 2 sublateral and 1 ventral. Setae on non-annulated tail region. Differentiated from other 2 stages by 4 to 6 CAT, 3 rows of PAT.

32. *Cygnonema,* Gr. n., swan nematode.

Cygnonema steineri[33] n. sp.

(Figs. 167 to 175)

Measurements (15 ♀♀): L = 2.4 (1.7-3.0) mm; b = 4.5 (3.9-5.5); c = 24.9 (16.3-37.0); V = 68 (65-73)%; CAT (Ant) = 20 (16-25) μm; CAT (Post) = 36 (33-39) μm; SS = 10-67 μm; first SlAT = 50 (44-55) μm; last SlAT = 26 (21-31) μm; first SvAT = 34 (31-38) μm; last SvAT = 17 (15-24) μm; No SlAT = 21 (15-25); No SvAT = 12 (10-15); Non-ann Term to Tail Length = 35 (23-49)%; T/ABD = 3.5 (2.6-4.3).

(8 ♂♂): L = 2.5 (2.1-3.1) mm; a = 28.3 (22.2-38.5); b = 5.0 (4.3-6.3); c = 18.1 (15.3-21.2); CAT (Ant) = 23 (21-26) μm; CAT (Post) = 40 (37-44) μm; SS = 9-67 μm; first SlAT = 53 (52-56) μm; last SlAT = 30 (25-33) μm; first SvAT = 36 (32-41) μm; last SvAT = 19 (16-25) μm; No SlAT = 19 (16-25); No SvAT = 10 (8-13); Non-ann Term to Tail Length = 26 (19-33)%; T/ABD = 3.6 (2.3-5.1); Spic = 83 (71-98) μm; Gub = 21 (17-24) μm.

(Holotype ♂): L = 2.2 mm; a = 27.5; b = 4.6; c = 18.3; CAT (Ant) = 25 μm; CAT (Post) = 38 μm; SS = 12-66 μm; first SlAT = 52 μm; last SlAT = 31 μm; first SvAT = 41 μm; last SvAT = 21 μm; No SlAT = 17; No SvAT = 9; Non-ann Term to Tail Length = 31%; T/ABD = 3.2; Spic = 82 μm; Gub = 17 μm.

Male Holotype.—(Figures in parentheses refer to range within species.) Amphids conspicuous, elongate loop-shaped, with elongated dorsal arm (fig. 169). Ten CAT, paired (usually paired); in 4 longitudinal rows, 2 sublateral and 2 subdorsal, and 4 (4 to 6) transverse rows posterior to rostrum (1 specimen with 1 tube on rostrum). Anterior CAT just posterior to rostrum, and posterior tube 2 rostral widths (2 to 3 rostral widths) or 28 annules (26 to 36 annules) posterior to rostrum. Longest SS on mid-body region. Eight rows of SS on esophageal and mid-body regions, 4 sublateral, 2 subventral and 2 subdorsal; 4 sublateral rows on annulated tail region. Caudal glands extend anterior to anus 2.5 (2.5 to 3.6) times ABD. Eight (5 to 8) pairs of stout, broad-based uniformly tapered anal setae, 15 to 18 μm (11 to 20 μm) long, 4 (3 to 4) pairs anterior to and 4 (2 to 4) posterior to anus. Short anal flap. One stout, lateral pair of setae (lateral to subventral, usually lateral) about 50% on non-annulated tail region, position measured from last tail annule to tail tip.

Females.—Similar to males. Amphids inconspicuous, tubular-shaped. Ten to 13 CAT, usually 10, paired, in 3 to 5 transverse rows, generally posterior to rostrum. Anterior CAT usually just posterior to rostrum, 1 specimen with 3 tubes on rostrum; posterior tube 2 ½ to 3 rostral widths or 27 to 37 annules posterior to rostrum. Esophagus terminal swelling slightly longer than in males. Paravulval setae typical of genus (fig. 170), 5 to 9 μm long; usually 2 pairs anterior to vulva; and 2 to 3 pairs, usually 2, posterior to vulva. Longest SS on mid-body region. Caudal glands extend anterior to anus 3.1 to 4.1 times ABD. One stout, lateral to subventral, usually subventral, pair of setae about 50% on non-annulated tail region.

Second-Stage Larvae.—L = 0.7 to 1.0 mm. Similar to adults. Amphids tubular-shaped. CAT typical of genus, tube less than 1 rostral width or 1 to 4 annules posterior to rostrum. Four sublateral rows of SS on esophageal, mid-body, and annulated tail region. SlAT typical of genus. One pair of stout, subventral to lateral, usually subventral, pair of setae usually just posterior to last complete tail annule sometimes just anterior to last complete tail annule on non-annulated tail region.

Third-Stage Larvae.—L = 1.1 to 1.4 mm. Similar to adults. Amphids tubular-shaped. CAT typical of genus, posterior tube 1 to 1 ½ rostral widths or 9 to 18 annules posterior to rostrum. Six rows of SS on esophageal region, 4 sublateral, 1 dorsal and 1 ventral. Five rows of SS on mid-body region, 4 sublateral and 1 dorsal; 4 sublateral rows on annulated tail region. Six to 9 SlAT, usually not paired, if paired, 7 pairs. Short anal flap present or absent. One stout, pair of setae, usually subventral just posterior to last complete tail annule on non-annulated tail region.

Fourth-Stage Larvae.—L = 1.7 to 2.2 mm. Similar to adults. Amphids tubular-shaped. Usually 4 CAT, some with 5 or 6; anterior tube usually just posterior to rostrum and posterior tube 2 to 3 rostral widths or 22 to 33 annules posterior to rostrum. Six rows of SS on esophageal region, 4 sublateral, 1 dorsal and 1 ventral. Six rows of SS on mid-body region, 4 sublateral and 2 subdorsal; 4 sublateral rows on annulated tail region. Nine to 13 SlAT and 6 to 8 VAT. Setae on non-annulated

33. Named in honor of G. Steiner.

tail region present or absent; usually 1 subdorsal or subventral pair about 50%; sometimes 1 stout, subventral pair just posterior to last complete tail annule.

Holotype (♂).—Collected December 10, 1969 by R. W. Timm and D. R. Viglierchio. U. S. National Museum of Natural History, Smithsonian Institution, Washington, D. C., No. 52006.

Paratypes.—3 ♀♀, 2 ♂♂. Same data as holotype. One female and 1 male deposited at U. S. National Museum of Natural History, Smithsonian Institution, Washington, D. C.; and 2 females and 1 male at University of California, Davis (UCNC 1534 to 1535).

Larval Stages.—2 second, 4 third, and 2 fourth. Same data as holotype. Deposited in same nematode collections as paratypes.

Type Habitat.—Marine, collected at 457 meters.

Type Locality.—Opposite Hut Point, McMurdo Sound, Antarctica.

Distribution.—Duke Ernst Bay, Gauss Station, McMurdo Sound, and Scott Base, Antarctica (other locations in Antarctica—on the slides of G. Steiner).

Diagnosis.—Differ by number of CAT, SlAT, and SvAT; location of posterior CAT; and amphid shape.

All larval stages differ by location of CAT, and amphid shape. Fourth-stage larvae differ by number of SlAT and VAT.

Discussion.—Specimens of *C. steineri* were present on some slides of G. Steiner's that are deposited in the USDA Nematode Collection, Beltsville, Maryland. These specimens were collected on the Deutschen Südpolar-Expedition, 1901 to 1903. We were able to remount many of these specimens, whose condition ranged from good to poor.

Genus *Dracognomus*[34] n. gen.

Diagnosis: Dracognominae. Nematodes 0.3 to 0.6 mm long. Body almost "epsilonematid" in shape. Esophageal region slightly swollen, swollen more dorsally. Swollen esophageal region followed by slender region, mid-body slightly to obviously swollen, followed by slender region, remainder of body to anus slightly swollen (fig. 176). Greatest width in both sexes usually at mid-body. Rostrum ventrally directed, dorsal side oblique; broadly rounded anteriorly; stoma opening slightly ventral (fig. 178). Rostrum ornamented with conspicuous punctations or obscure subcuticular granulations, setae present. Amphids inconspicuous, inverted "U" shape (staple-shaped); dorso-laterally at base of rostrum extends into first few body annules (fig. 178). Most adhesion tubes modified (Mod-AT); (see Morphology, p. 14, and fig. 177). Eight Mod-CAT, 2 sublateral and 2 subdorsal pairs, in 2 transverse rows on anterior $1/2$ of rostrum or less than 1 rostral width posterior to rostrum. Buccal cavity well developed, cylindrical, with conspicuous dorsal tooth (fig. 178). Esophagus with small median swelling, posterior bulb with conspicuous valve (fig. 176). Cuticle not unusually thick. Annules with or without obscure subcuticular granulation; sometimes with break in annulation forming lateral line, when present, line usually extends posteriorly from just anterior to swollen mid-body. Distinct or indistinct rows of SS, some rows with alternating long and short setae. Paravulval setae present. Females with subventral Mod-AT on esophageal region, or sublateral Mod-AT on mid-body region. PAT in 4 longitudinal rows, 2 sublateral and 2 subventral; with or without tubes posterior to anus; number of tubes variable between species. PAT either all Mod-AT, or with both normal tubes and Mod-AT (fig. 179). Male anal setae absent. Anal flap absent. Setae on non-annulated tail region. Males with subventral cuticularized protuberants on non-annulated tail region, absent on females.

Type Species: *Dracognomus marioni* n. sp.

Differs by Mod-AT; well-developed, cylindrical, buccal cavity with conspicuous dorsal tooth; esophagus with small median swelling, and posterior esophageal bulb with conspicuous cuticularized valve.

Larval Stages: No first-, second-, or third-stage larvae available in *Dracognomus*. Fourth-stage larvae 0.3 to 0.4 mm long. Similar to adults. Swollen body regions not as prominent as in adults.

---

34. *Dracognomus*, NL. m., little dragon.

114 University of California Publications in Zoology

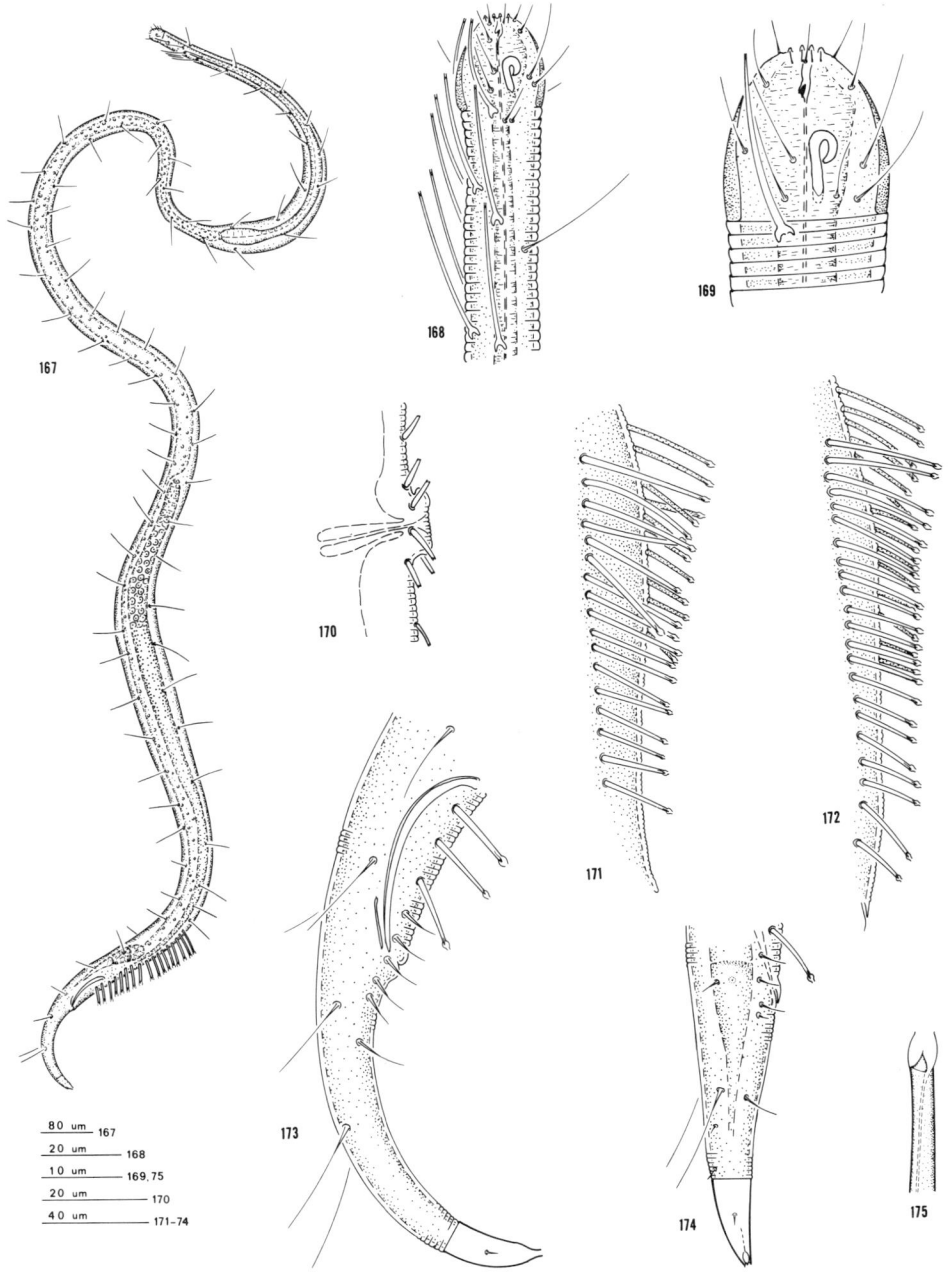

28. Figs. 167–175: *Cygnonema steineri* n. sp. 167) Male, full length[a], SvAT and correct number of CAT not illustrated; 168) Male, head and CAT region, correct number of CAT not illustrated; 169) Male, head; 170) Female, vulval region, note tubular paravulval setae; 171) Male, PAT; 172) Female, PAT; 173) Male, anal region and tail; 174) Female, tail; 175) Male, posterior adhesion tube, note bell-shaped structure with internal tongue-like triangular structure.

a. Total number of setae on non-annulated tail not illustrated.

Rostrum not as oblique dorsally and anterior not as broadly rounded as in adults. Amphids inverted "U" shaped. Adhesion tubes either all Mod-AT, or with both normal and Mod-AT. Four Mod-CAT, 2 sublateral and 2 subdorsal, in 2 transverse rows, either on anterior $1/2$ rostrum or less than 1 rostral width posterior to rostrum. SS usually in distinct rows on body, some rows with alternating long and short setae. PAT in 4 longitudinal rows, 2 sublateral and 2 subventral; tubes in sublateral rows Mod-AT or normal, tubes in subventral rows Mod-AT and/or normal (see Discussion). Number of tubes variable between species. Anal flap absent. With or without setae on non-annulated tail region. Differentiated by 4 Mod-CAT, and PAT in 4 rows.

Discussion: All fourth-stage larvae examined in this superfamily with 3 longitudinal rows of PAT, had normal adhesion tubes. *Dracognomus* is the only genus in this investigation that had fourth-stage larvae with 4 longitudinal rows of PAT. In this genus the adhesion tubes were all Mod-AT, except for 1 anterior pair in *D. simplex* n. comb. Because the 4 row condition present in *Dracognomus* is composed mainly of Mod-AT which differs structurally from the normal adhesion tubes, we did not consider this genus as normal when referring to fourth-stage larvae as having only 3 longitudinal rows of PAT.

With further study, this genus may very well prove to be a link between the Draconematoidea Filipjev, 1918 and Epsilonematoidea n. grad. Structure of the Mod-AT and their location on the females in this genus seem to have homologies both to the "adhesion tubes" characteristic of Draconematoidea and "stilt setae" of Epsilonematoidea.

Key to Species of *Dracognomus*

1. PAT all Mod-AT (fig. 176). One[35] to 3 Mod-SlAT posterior to anus. Mod-CAT on anterior $1/2$ of rostrum 2
PAT of both types, normal and Mod-AT (fig. 179). All Mod-SlAT anterior to anus. Mod-CAT less than 1 rostral width or 3 to 4 annules posterior to rostrum. . . . .
. . . . *simplex* (Gerlach, 1954) n. comb. (p. 118)
2. Males with 8 to 9 Mod-SlAT; and 2 pairs of subventral cuticularized protuberants on non-annulated tail region. Females 0.5 to 0.6 mm long; posterior subventral Mod-AT on esophageal region adjacent to or near posterior esophageal bulb; 1 pair of paravulval setae anterior to vulva and 1 pair posterior. *marioni* n. sp. (p. 115)
Males with 6 Mod-SlAT; and 1 pair of subventral cuticularized protuberants on non-annulated tail region. Females 0.3 mm long; posterior subventral Mod-AT on esophageal region posterior to posterior esophageal bulb; 1 paravulval seta anterior to vulva on right side of body, and 1 paravulval seta posterior to vulva on left . . . . *notohalensis* n. sp. (p. 119)

Dracognomus marioni[36] n. sp.
(Figs. 176 to 178)

Measurements (4 ♀♀): L = 0.6 (0.5–0.6) mm; b = 7.4 (6.7–8.3); c = 13.3 (11.8–15.0); V = 51 (48–53)%; Mod-CAT = 15 (14–16) µm; SS = 11–40 µm; first Mod-SlAT = 12 (11–14) µm; last Mod-SlAT = 14 (12–15) µm; first Mod-SvAT = 7 (6–7) µm; last Mod-SvAT = 10 (10–11) µm; No Mod-SlAT = 11 (9–12); No Mod-SvAT = 8 (7–8); Non-ann Term to Tail Length = 50 (46–52)%; T/ABD = 2.4 (2.3–2.4).

(4 ♂♂): L = 0.5 (0.4–0.6) mm; a = 13.0 (11.0–16.3); b = 7.9 (5.9–9.2); c = 10.1 (9.2–11.0); Mod-CAT = 13 (12–14) µm; SS = 11–33 µm; first Mod-SlAT = 12 (12–14) µm; last Mod-SlAT =

---

35. All counts and measurements on right side of body.
36. Named in honor of A. F. Marion.

14 (13–15) μm; first Mod-SvAT = 6 (6–7) μm; last Mod-SvAT = 10 (8–12) μm; No Mod-SlAT = 9 (8–9); No Mod-SvAT = 8 (6–8); Non-ann Term to Tail Length = 42 (39–45)%; T/ABD = 2.4 (2.3–2.4); Spic = 32 (30–33) μm; Gub = 18 (17–19) μm.

(Holotype ♂): L = 0.6 mm; a = 16.3; b = 8.9; c = 10.3; Mod-CAT = 14 μm; SS = 11–33 μm; first Mod-SlAT = 12 μm; last Mod-SlAT = 15 μm; first Mod-SvAT = 6 μm; last Mod-SvAT = 12 μm; No Mod-SlAT = 9; No Mod-SvAT = 8; Non-ann Term to Tail Length = 43%; T/ABD = 2.4; Spic = 30 μm; Gub = 18 μm.

Male Holotype.—(Figures in parentheses refer to range within species.) Rostrum ornamented with punctations. Amphids typical of genus. Mod-CAT on anterior $1/2$ rostrum. Annules on esophageal region with subcuticular granulation (absent on some specimens), remaining annulation without ornamentation. Short lateral line (absent on some specimens), just anterior to and extending posteriorly to swollen mid-body. Longest SS on esophageal region. SS not in distinct rows on esophageal region, probably 8, 4 sublateral, 2 subdorsal and 2 subventral. Eight rows of SS on mid-body region, 4 sublateral, 2 subventral and 2 subdorsal; 4 sublateral rows on annulated tail region. Only Mod-CAT and PAT present. PAT all Mod-AT. Long and short setae intermingled with Mod-SlAT. Two (2 to 3) long setae with Mod-SlAT, not alternating with tubes, setae about equal in length to adjacent tubes. One of the Mod-SlAT adjacent to and 1 posterior to anus (some specimens with 2 tubes posterior to anus on left side of body). Mod-SlAT with 3 long and 6 short tubes. Caudal glands extend anterior to anus 1.8 (1.7 to 2.2) times ABD. Two pairs of subventral cuticularized protuberants on non-annulated tail region, position measured from last complete tail annule to tail tip; 1 pair about 13% (13% to 25%); and 1 pair about 50% (50% to 66%), (fig. 176). One pair (1 to 2) latero-subventral setae (lateral to subventral, usually subventral) on non-annulated tail region; 1 pair just anterior to first pair of protuberants (some specimens just posterior to last complete tail annule, and some with 1 pair of setae just posterior to first pair).

Females.—Similar to males. Two subventral rows of 4 to 5 Mod-AT on esophageal region (fig. 178), tubes 6 to 10 μm long; rows beginning just posterior to rostrum, extending posteriorly to posterior esophageal bulb, 1 specimen with 1 tube less than $1/4$ body width posterior to bulb. Annules from 1 ABD anterior to anus to last tail annule with subcuticular granulation. Two subventral pairs of paravulval setae, tubes 6 to 9 μm long, 1 pair anterior and 1 posterior to vulva. Four inconspicuous ventral setae just anterior to Mod-SvAT. Longest SS on esophageal region and opposite PAT. Subventral rows of SS present only posterior to vulva on mid-body region, without setae intermingled with Mod-SlAT. Mod-SlAT with 3 to 6 long and 5 to 9 short tubes. Two to 3 of the Mod-SlAT posterior to anus (fig. 177). Caudal glands extend anterior to anus 1.3 to 2.1 times ABD. One subventral pair of setae, present or absent (usually absent), 2 to 4 annule widths posterior to last complete tail annule on non-annulated tail region.

First-, Second-, and Third-Stage Larvae.—Not observed.

Fourth-Stage Larvae.—L = 0.3 to 0.4 mm. Similar to adults. Greatest body width at esophageal and mid-body regions. Mod-CAT typical of genus, on anterior $1/2$ of rostrum. SS not in distinct rows on esophageal region, probably 7, 4 sublateral, 2 subdorsal and 1 ventral. Six rows of SS on mid-body region, 4 sublateral, 1 dorsal and 1 ventral; 4 sublateral rows on annulated tail region. Only Mod-CAT and PAT present. PAT all Mod-AT. Four pairs of Mod-SlAT, with 1 of the Mod-SlAT posterior to anus. Four to 6 unpaired Mod-SvAT. One stout, lateral to subventral pair of setae, present or absent, just posterior to last complete tail annule on non-annulated tail region.

Holotype (♂).—Collected August 14, 1969 by R. T. DeNeve. Catalogue No. UCNC 1438, University of California, Davis.

Paratype.—5 ♀♀, 7 ♂♂. Same data as holotype. Deposited at University of California, Davis (UCNC 1539 to 1540); and USDA Nematode Collection, Beltsville, Maryland.

Larval Stages.—2 fourth stage. Same data as holotype. Deposited at University of California, Davis.

Type Habitat.—Marine, collected from sand on sand bar.

Type Locality.—Boca Raton, Florida, USA; in Palm Beach County on east coast.

Distribution.—Same as type locality.

Diagnosis.—Most closely resembles *D. notohalensis* n. sp. differs by greater number of Mod-SlAT and subventral cuticularized protuberants on non-annulated tail region in males; females by larger size, 2 pairs of paravulval setae, and posterior tube of subventral Mod-AT on esophageal region ad-

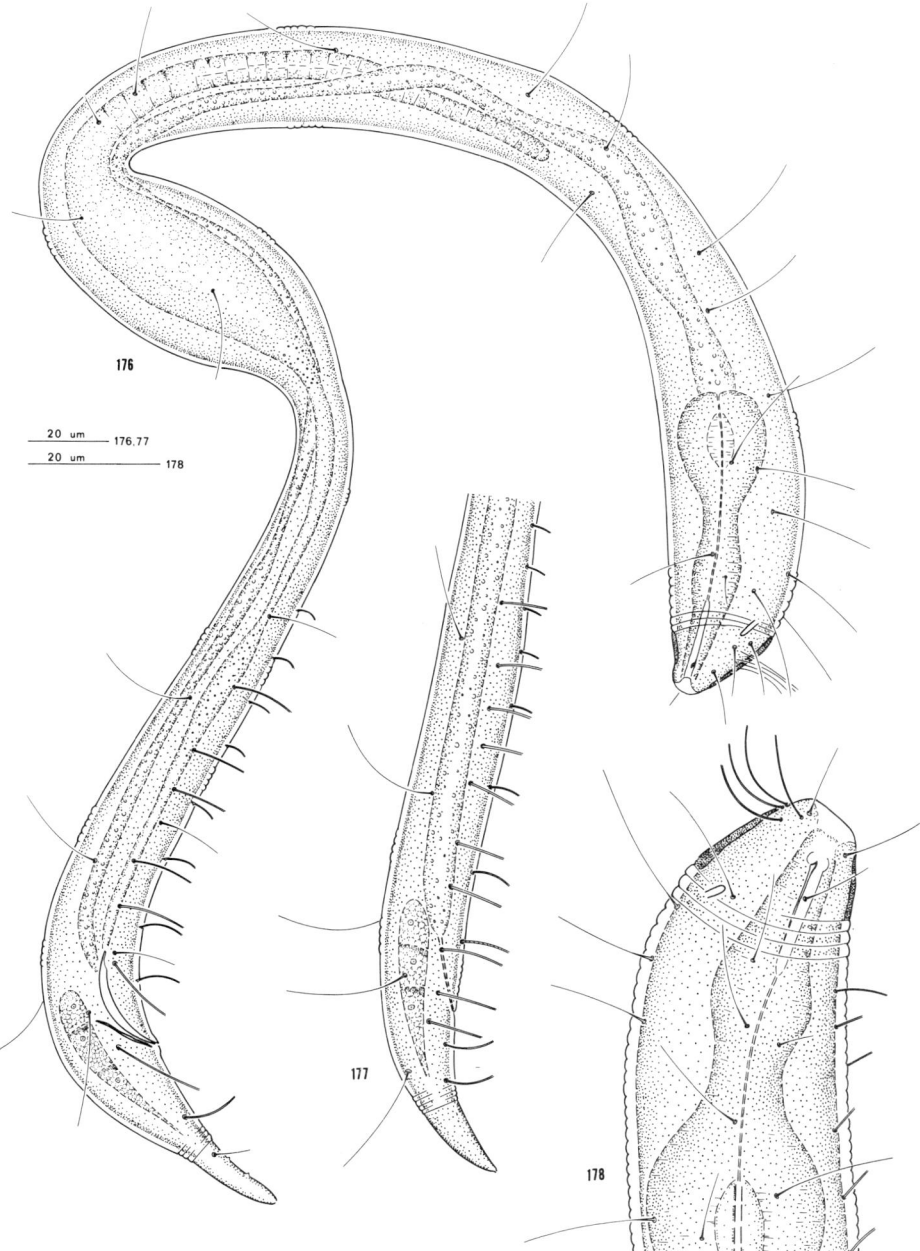

29. Figs. 176-178: *Dracognomus marioni* n. sp. 176) Male, full length[a]; 177) Female, posterior body region; 178) Female, head and esophageal region.

a. Short SS with Mod-SlAT not illustrated.

jacent to posterior esophageal bulb. Differs from *D. simplex* n. comb. by only Mod-AT, Mod-CAT on anterior $1/2$ of rostrum, and some Mod-SlAT posterior to anus.

Fourth-stage larvae differ from fourth-stage *D. simplex* n. comb. by only Mod-AT and Mod-CAT, on anterior $1/2$ of rostrum.

Dracognomus simplex (Gerlach, 1954) n. comb.

(Figs. 179, 182)

Syn: *Drepanonema simplex* Gerlach, 1954 n. syn.

Measurements (10 ♀♀): L = 0.3 (0.3–0.4) mm; b = 10.0 (7.7–15.0); c = 6.5 (5.0–8.8); V = 47 (45–50)%; Mod-CAT = 15 (12–17) μm; SS = 5–30 μm; first SlAT = 19 (16–21) μm; last SlAT = 23 (20–25) μm; first SvAT = 15 (13–16) μm; last SvAT = 10 (9–12) μm; first Mod-SvAT = 11 (10–15) μm; last Mod-SvAT = 15 (13–17) μm; No SlAT = 4; No SvAT = 2 (2–3); No Mod-SvAT = 3 (3–4); Non-ann Term to Tail Length = 70 (67–71)%; T/ABD = 4.5 (4.1–4.6).

Females Emended.—Esophageal region with slight ventral swelling (fig. 179). Rostrum with obscure subcuticular granulation. Mod-CAT less than 1 rostral width or 3 to 4 annules posterior to rostrum. Two sublateral rows of 7 to 9 Mod-AT, tubes 6 to 15 μm long, rows beginning just posterior to the esophageal region extending posteriorly almost to SlAT (fig. 179). Annules with obscure subcuticular granulation, most prominent just anterior to anus. Two subventral pairs of paravulval setae, 6 to 7 μm long, 1 pair anterior to vulva and 1 pair posterior. Without conspicuous ventral setae just anterior to SvAT. Longest SS opposite PAT and on annulated tail region. SS not in distinct rows on esophageal region, probably 8, 4 sublateral, 2 subdorsal and 2 subventral. Seven rows of SS on mid-body region, 4 sublateral, 2 subdorsal and 1 ventral; 4 sublateral rows on annulated tail region. PAT of both types, Mod-AT and normal (fig. 179). Without setae intermingled with SlAT. SlAT all normal tubes, without SlAT posterior to anus. SlAT with 2 long and 2 short tubes. Subventral PAT of both types, anterior tubes normal, posterior tubes Mod-AT. Caudal glands extend anterior to anus 1.7 to 2.5 times ABD. One lateral to subventral pair of setae on non-annulated tail region, about 25%, position measured from last complete tail annule to tail tip.

Males, First-, Second-, and Third-Stage Larvae.—Not observed.

Fourth-Stage Larvae Emended.—L = 0.3 mm. Similar to female. Greatest body width at esophageal and mid-body regions. Mod-CAT less than 1 rostral width posterior to rostrum. Four sublateral rows of SS on esophageal region. Five rows of SS on mid-body region, 4 sublateral and 1 dorsal. Two dorso-lateral rows of SS on annulated tail region. Only Mod-CAT and PAT present. PAT of both types, Mod-AT and normal. All SlAT normal tubes, without SlAT posterior to anus. Four pairs of SlAT, middle pairs shorter. First anterior SvAT normal, remaining 5 Mod-SvAT. Non-annulated tail region without subventral cuticularized protuberants or setae.

Type Habitat.—Marine, in subterranean water along shore.

Type Locality.—Banyuls and Cannes, France, Mediterranean Sea.

Distribution.—Banyuls and Cannes, France, Mediterranean Sea; and Terrenia, Italy, The Ligurian Sea.

Diagnosis.—Differs from other known females of *Dracognomus* by both Mod-AT and normal adhesion tubes, Mod-CAT just posterior to rostrum, and absence of Mod-SlAT posterior to anus.

Fourth-stage larvae differ from fourth-stage *D. marioni* n. sp. by both types of SvAT, normal and Mod-AT.

Type material of this species was not available for study. This species was easily identifiable by comparing specimens collected near the type locality with the original description.

Dracognomus notohalensis[37] n. sp.

(Figs. 180 to 181, 183)

Measurements (1 ♂): L = 0.3 mm; a = 14.0; b = 7.0; c = 7.0; Mod-CAT = 20 μm; SS = 8–33 μm; first Mod-SlAT = 14 μm; last Mod-SlAT = 13 μm; first Mod-SvAT = 8 μm; last Mod-SvAT = 12 μm; No Mod-SlAT = 6; No Mod-SvAT = 6; Non-ann Term to Tail Length = 42%; T/ABD = 2.4; Spic = 24 μm; Gub = 9 μm.

(Holotype ♀): L = 0.3 mm; b = 6.8; c = 9.0; V = 48%; Mod-CAT = 17 μm; SS = 9–36 μm; first Mod-SlAT = 14 μm; last Mod-SlAT = 14 μm; first Mod-SvAT = 9 μm; last Mod-SvAT = 9 μm; No Mod-SlAT = 10; No Mod-SvAT = 7; Non-ann Term to Tail Length = 49%; T/ABD = 2.6.

Female Holotype.—Rostrum ornamented with punctations. Mod-CAT on anterior $1/2$ of rostrum. Two subventral rows of 4 Mod-AT on esophageal region, tubes 6 μm long; rows beginning just posterior to rostrum extending posteriorly to just posterior to esophageal bulb (fig. 181). Annules with obscure granulation; short lateral line beginning just anterior to and extending posteriorly to swollen mid-body. Single paravulval seta on each side of body; seta on right side of body anterior to vulva, 6 μm long; seta on left posterior to vulva, 4 μm long. Five conspicuous, stout, slightly broad-based, ventral setae just anterior to Mod-SvAT. Longest SS on esophageal region. SS on esophageal region not in distinct rows, probably 6, 4 sublateral and 2 subdorsal. Eight rows of SS on mid-body region, 4 sublateral, 2 subdorsal and 2 subventral; 4 sublateral rows on annulated tail region. Without setae intermingled with Mod-SlAT. PAT all Mod-AT. Mod-SlAT with 3 long and 7 short tubes. Two of the Mod-SlAT posterior to anus. Caudal glands extend anterior to anus 1.5 times ABD. One lateral pair of setae just posterior to last complete tail annule on non-annulated tail region.

Male.—Similar to female. Lateral line beginning just anterior to swollen mid-body extending posteriorly through mid-body. Longest SS on esophageal region. Five conspicuous, stout, slightly broad-based, ventral setae just anterior to Mod-SvAT. SS not in distinct rows on esophageal region, probably 7, 4 sublateral, 2 subdorsal and 1 ventral. Eight rows of SS on mid-body region, 4 sublateral, 2 subdorsal and 2 subventral; and 4 sublateral rows on annulated tail region. Only Mod-CAT and PAT present. PAT all Mod-AT. Long and short setae intermingled with Mod-SlAT. Three long setae with Mod-SlAT, not alternating with tubes, setae about equal in length to adjacent tubes. Mod-SlAT with 3 long and 3 short tubes. One of the Mod-SlAT posterior to anus. Caudal glands extend anterior to anus 1.1 times ABD. One subventral pair of cuticularized protuberants on non-annulated tail region (fig. 183), 33% to 50%, position measured from last complete tail annule to tail tip. One subventral pair of setae just posterior to last complete tail annule on non-annulated tail region.

Larval Stages.—Not observed.

Holotype (♀).—Collected November 26, 1969 by R. W. Timm. Catalogue No. UCNC 1439, University of California, Davis.

Paratype.—(♂). Same data as holotype. Deposited at University of California, Davis (UCNC 1542).

Type Habitat.—Marine, associated with coral on coral reef.

Type Locality.—Pago Pago Island, American Samoa, Samoa Islands.

Distribution.—Same as type locality.

Diagnosis.—Most closely resembles *D. marioni* n. sp. differs by fewer Mod-SlAT and subventral cuticularized protuberants on non-annulated tail region in males; females by 1 paravulval seta on each side of body, posterior tube of subventral Mod-AT on esophageal region just posterior to posterior esophageal bulb, and smaller size. Differs from *D. simplex* n. comb. by only Mod-AT, Mod-CAT on anterior $1/2$ of rostrum, and some Mod-SlAT posterior to anus.

Genus *Notochaetosoma* Irwin-Smith, 1918

Diagnosis Emended: Notochaetosomatinae. Nematodes 0.9 to 1.6 mm long. Body without conspicuous swollen regions, width nearly equal entire body. Esophageal region slightly swollen, greatest width at mid-body (fig. 184). Rostrum broadly rounded anteriorly, cuticle 4 μm thick, with obscure

---

37. *notohalensis*, south sea.

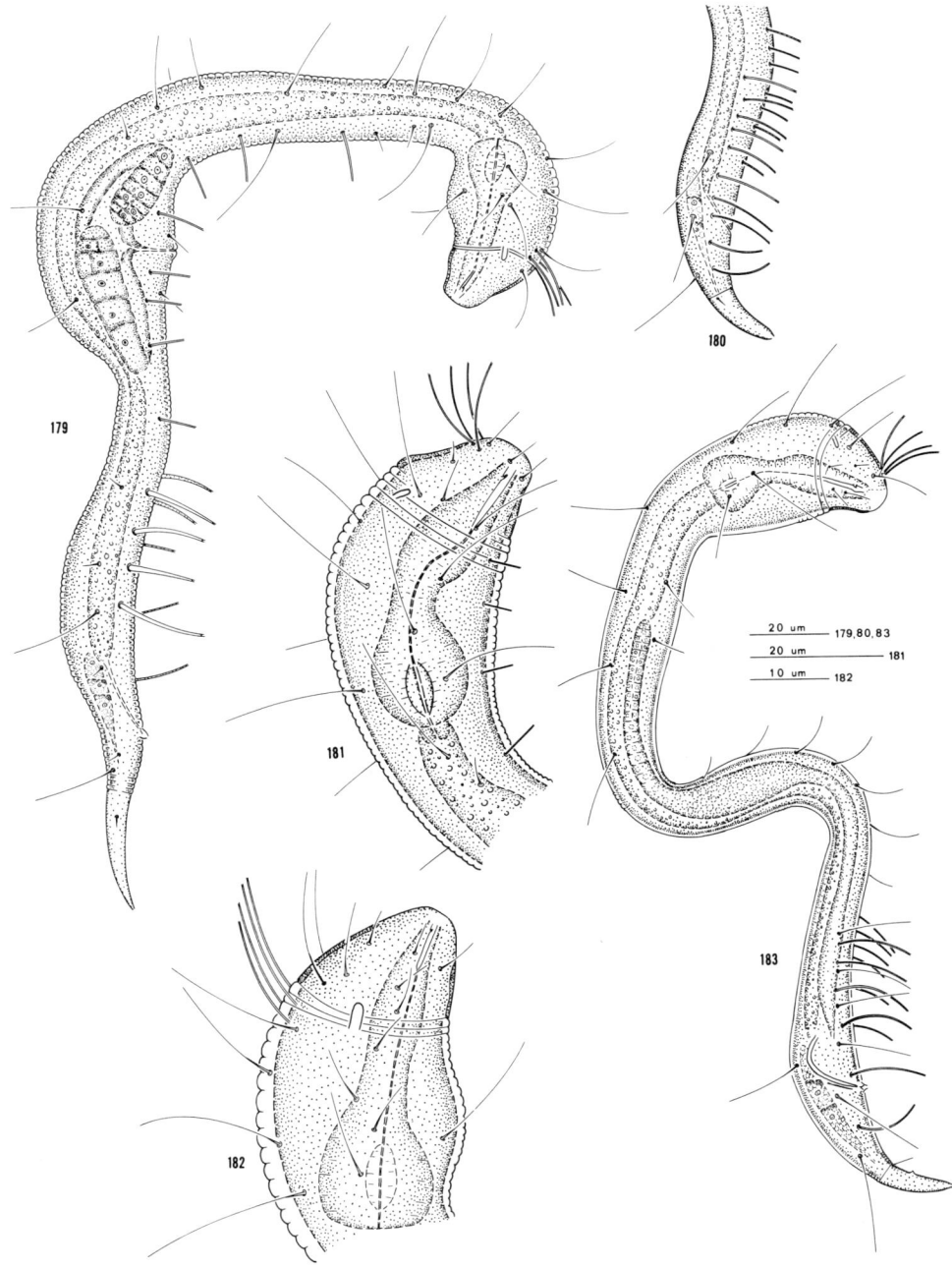

30. Figs. 179, 182: *Dracognomus simplex* (Gerlach, 1954) n. comb. 179) Female, full length, note Mod-AT and normal tubes; 182) Female, head and esophageal region. Figs. 180–181, 183: *Dracognomus notohalensis* n. sp. 180) Female, posterior body region; 181) Female, head and esophageal region; 183) Male, full length[a].

a. Short SS with Mod-SlAT not illustrated.

markings, setae present. Amphids conspicuous, dorso-laterally on rostrum; male amphids loop-shaped (horseshoe-shaped); female amphids usually unispiral, sometimes loop-shaped. Eight CAT, paired; in 2 transverse rows, anterior row on rostral base, posterior row less than 1 rostral width or 2 to 3 annules posterior to rostrum. Buccal cavity weakly developed, collapsed, unarmed. Esophagus with cylindrical corpus, and slight terminal swelling without cuticularized valve. Cuticle very thick, thickness may obscure internal structures. Annules with subcuticular granulation, most prominent on anterior and posterior body regions; a few vacuoles present. Tail annules without ornamentation. SS in distinct rows on body, some rows with alternating long and short setae. Paravulval setae present. Four longitudinal rows of PAT, 2 sublateral and 2 subventral. Males with uniformly tapered anal setae. Anal flap absent. Anterior $2/3$ of non-annulated tail region with prominent punctations, posterior $1/3$ without or very few punctations; setae present. Cuticularized protuberants absent on non-annulated tail region.

Type Species: *Notochaetosoma tenax* Irwin-Smith, 1918

Differs from *Dracogalerus* n. gen. by posterior row of CAT posterior to rostrum, and rostral cuticle 4 μm thick.

Larval Stages: No first-stage larvae available in *Notochaetosoma*. Second-stage larvae 0.5 to 0.6 mm long. Similar to adults. Body without swollen regions. Amphids conspicuous, dorso-laterally on rostrum; usually loop-shaped (horseshoe-shaped), some specimens with ventral arm curved dorsally almost forming unispiral. Single dorsal CAT, less than 1 rostral width or 1 to 2 annules posterior to rostrum. Annules with obscure subcuticular granulation, most prominent on anterior part of body. SS in distinct rows on body. Two pairs of SlAT, in 2 longitudinal rows. Anal flap absent. Differentiated from other 2 stages by 1 CAT, 2 pairs of SlAT in 2 rows.

Third-stage larvae 0.5 to 0.7 mm long. Similar to adults. Greatest body width at esophageal and mid-body regions. Amphids conspicuous, dorso-laterally on rostrum; usually loop-shaped (horseshoe-shaped), sometimes unispiral. Three CAT, 2 subdorsal and 1 dorsal, posterior tube less than 1 rostral width or 1 to 3 annules posterior to rostrum. Annules on esophageal region with obscure subcuticular granulation, posterior body region with obscure granulation and vacuoles. SS in distinct rows on body, some rows with alternating long and short setae. SlAT in 2 longitudinal rows, usually 6 pairs, if unpaired, 6 on right and 5 on left side of body. Anal flap absent. Differentiated from other 2 stages by 3 CAT, 2 rows of SlAT.

Fourth-stage larvae 0.8 to 1.1 mm long. Similar to adults. Greatest body width at esophageal and mid-body regions. Amphids conspicuous, dorso-laterally on rostrum; usually unispiral, sometimes loop-shaped (horseshoe-shaped). Four CAT, 2 sublateral and 2 subdorsal, just posterior to rostrum or on base of rostrum. If CAT on rostrum posterior tube just anterior to first body annule, if posterior to rostrum posterior tube less than 1 rostral width or 1 to 2 annules posterior to rostrum. Annules with subcuticular granulation, most prominent on anterior part of body. SS in distinct rows on body, some rows with alternating long and short setae. PAT in 3 longitudinal rows, 2 sublateral and 1 ventral. Anal flap absent. Differentiated from other 2 stages by 4 CAT, 3 rows of PAT.

Discussion: In 1963, D. G. Murphy described *Notochaetosoma costeriata,* which was correctly transferred to the genus *Epsilonema* (Family: Epsilonematidae), by Lorenzen in 1973.

Notochaetosoma tenax Irwin-Smith, 1918

(Figs. 184 to 190)

Measurements (8 ♀♀): L = 1.3 (0.9–1.6) mm; b = 11.3 (8.5–13.0); c = 14.8 (10.6–17.2); V = 42 (40–44)%; CAT = 24 (22–25) μm; SS = 11–34 μm; first SlAT = 24 (22–27) μm; last SlAT = 26 (24–28) μm; first SvAT = 23 (21–25) μm; last SvAT = 20 (15–22) μm; No SlAT = 18 (16–21); No SvAT = 19 (16–22); Non-ann Term to Tail Length = 59 (52–69)%; T/ABD = 3.2 (2.7–3.7).

(3 ♂♂): L = 1.3 (1.1–1.4) mm; a = 17.6 (16.3–18.9); b = 11.1 (10.8–11.3); c = 13.0 (12.4–13.6); CAT = 24 (23–25) μm; SS = 9–34 μm; first SlAT = 25 (24–27) μm; last SlAT = 25 (24–27) μm; first SvAT = 19 (15–24) μm; last SvAT = 21 (20–22) μm; No SlAT = 10 (8–12); No SvAT = 16 (12–20); Non-ann Term to Tail Length = 46 (45–47)%; T/ABD = 3.1 (2.7–3.3); Spic = 55 (50–59) μm; Gub = 15 (14–16) μm.

Males Emended.—Amphids usually loop-shaped (horseshoe-shaped); sometimes ventral arm curved dorsally, almost forming unispiral. CAT posterior row less than 1 rostral width or 2 annules posterior to rostrum. Longest SS on esophageal region. Eight rows of SS on esophageal and mid-body regions, 4 sublateral, 2 subdorsal and 2 subventral; 4 sublateral rows on annulated tail region. Long and short setae intermingled with SlAT. Six to 7 long setae with SlAT, not alternating with tubes, setae either equal in length or shorter than adjacent tubes. Caudal glands extend anterior to anus 2.7 to 3.3 times ABD. Two pairs of uniformly tapered anal setae, 11 to 12 μm long, 1 pair anterior and 1 posterior to anus. Four to 6 pairs of setae on non-annulated tail region, position measured from last complete tail annule to tail tip; 1 lateral to subventral pair, usually subventral, just posterior to last complete tail annule; 1 lateral to subventral pair about 25%; 2 lateral to subdorsal pairs, usually 1 about 50% and 1 66% to 75%; 2 subdorsal pairs, usually 1 about 25%, and 1 about 50%.

Females Emended.—Similar to males. Amphids usually unispiral sometimes loop-shaped (horseshoe-shaped). CAT posterior row less than 1 rostral width or 2 to 3 annules posterior to rostrum. Two subventral pairs of paravulval setae, 6 to 9 μm long, 1 pair anterior and 1 posterior to vulva. Vulva encircled with minute spine-like projections similar to *Paradraconema* n. gen., vulva may be obscured by thick cuticle. Longest SS on esophageal region. Short setae intermingled with SlAT. Caudal glands extend anterior to anus 2.4 to 3.3 times ABD. Four to 6 pairs of setae on non-annulated tail region; 1 short lateral to subventral pair about 50%; 1 short subventral pair (present or absent) just posterior to last complete tail annule, usually 1 short pair (either subdorsal, lateral or subventral) about 33%; 3 subdorsal pairs, 1 long pair just posterior to last complete tail annule or about 13% about equal in length to longest setae in sublateral row opposite PAT, 1 short pair about 33%, and 1 short pair about 50%.

Second-Stage Larvae.—L = 0.5 to 0.6 mm. Similar to adults. Rostrum ornamented with obscure subcuticular markings and a few vacuoles. Amphids and CAT typical of genus. Six rows of SS on esophageal region, 4 sublateral, 1 dorsal and 1 ventral. Five rows of SS on mid-body region, 4 sublateral and 1 dorsal; 4 sublateral rows on annulated tail region. SlAT typical of genus. Single dorsal seta on non-annulated tail region about 33%.

Third-Stage Larvae.—L = 0.5 to 0.7 mm. Similar to adults. Rostrum ornamented with obscure markings. Amphids and CAT typical of genus. Six rows of SS on esophageal region, 4 sublateral, 1 dorsal and 1 ventral. Five rows of SS on mid-body region, 4 sublateral and 1 dorsal; 4 sublateral rows on annulated tail region. SlAT typical of genus. One to 2 subdorsal pairs of setae on non-annulated tail region; usually 1 long pair just posterior to last complete tail annule about equal in length to longest setae in sublateral row opposite PAT, and 1 short pair 33% to 50%.

Fourth-Stage Larvae.—L = 0.8 to 1.1 mm. Similar to adults. Rostrum ornamented with vacuoles and punctations, most prominent at base. Amphids and CAT typical of genus. Eight rows of SS on esophageal region, 4 sublateral, 2 subdorsal and 2 subventral. Seven rows of SS on mid-body region, 4 sublateral, 2 subventral and 1 dorsal; 4 sublateral rows on annulated tail region. Eight to 11 SlAT, usually unpaired; 13 to 14 VAT. Two to 4 pairs of setae on non-annulated tail region; 1 short lateral to subventral pair, usually subventral, 25% to 33%; usually 1 short lateral to subdorsal pair 50% to 66%; 2 subdorsal pairs, 1 short pair 25% to 50%, and 1 long pair (present or absent) just posterior to last complete tail annule about equal in length to longest setae in sublateral row opposite the PAT.

Type Specimens (1 ♀, 1 ♂).—Collected in 1916, by V. A. Irwin-Smith. Australian Museum, Sydney, New South Wales, Australia, Nos. W456 and W457.

Type Habitat.—Marine, collected between high and low tidal zones.

Type Locality.—Vaucluse, Port Jackson, N. S. W., Australia.

Distribution.—General area of Port Jackson, Australia.

Diagnosis.—Differs by number of SlAT (♀♀ = 16 to 21; ♂♂ = 12 to 20), SvAT (♀♀ = 16 to 22; ♂♂ = 12 to 20), and setae on non-annulated tail region.

Type material of *N. tenax* was available for study, but these specimens were in poor condition. This species was easily identifiable by comparing specimens collected at the type locality to the original description and type specimens.

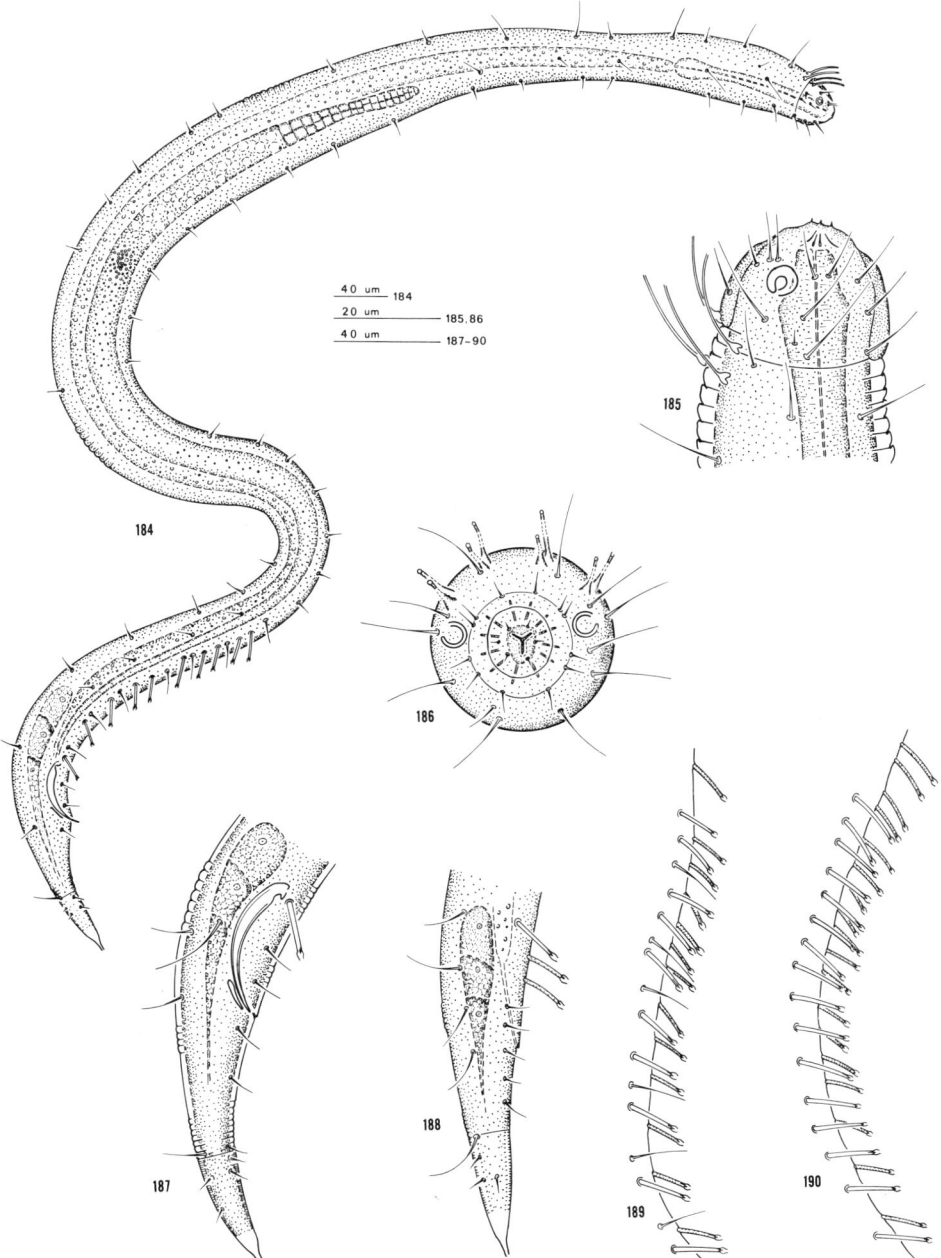

31. Figs. 184-190: *Notochaetosoma tenax* Irwin-Smith, 1918. 184) Male, full length[a], SvAT not illustrated; 185) Female, head; 186) Female, face view; 187) Male, tail; 188) Female, tail; 189) Male, PAT[a]; 190) Female, PAT[a].

a. Short SS with SlAT not illustrated.

Genus *Dracogalerus*[38] n. gen.

Diagnosis: Notochaetosomatinae. Nematodes 1.1 to 1.2 mm long. Body width without conspicuous swollen regions, nearly equal entire length, greatest width at mid-body (fig. 193). Rostrum either conoid or broadly rounded anteriorly, may or may not be expanded and set off from body; with or without ornamentation, rostral cuticle 7 to 12 μm thick, some rostral setae very stout about size of CAT. Amphids conspicuous, dorso-laterally on rostrum; loop-shaped (horseshoe-shaped), unispiral, partially doubled spiral ($1/8$ to $1/2$ spiral doubled). CAT paired, in 2 to 3 transverse rows on anterior $1/2$ of rostrum. Buccal cavity weak to moderately developed, unarmed. Esophagus corpus cylindrical, inconspicuous terminal swelling without cuticularized valve. Cuticle very thick, thickness may obscure internal structures. Annules with or without ornamentation. SS in distinct rows on body. Paravulval setae present. Four longitudinal rows of PAT, 2 sublateral and 2 subventral. Males with or without subventral corniform adhesion tubes posterior to last SvAT (see Morphology, p. 13, fig. 196), absent on females. Males with uniformly tapered anal setae. Anal flap absent. Male anal opening laterally bounded by oval area of smooth cuticle with obscure punctations, absent on females. Anterior $2/3$ to $4/5$ of non-annulated tail region with conspicuous punctations, posterior part without or very few punctations; setae present. Males with ventral protuberants on non-annulated tail region; absent on females. Setae present on non-annulated tail region.

Type Species: *Dracogalerus afrikaanus* n. sp.

Differs from *Notochaetosoma* Irwin-Smith, 1918 by CAT on anterior ½ of rostrum, and rostral cuticle 7 to 12 μm thick.

Larval Stages: No larval stages available in *Dracogalerus*.

Key to Species of *Dracogalerus*

1. Rostrum broadly rounded anteriorly; base of rostrum not expanded, rostrum not set off (fig. 192) 2
   Rostrum conoid shaped; base of rostrum expanded, wider than succeeding body width, rostrum set off (fig. 194). . . . . *afrikaanus* n. sp. (p. 124)
2. Ten CAT, in 3 transverse rows on rostrum . . . . *bastiani* n. sp. (p. 125)
   Eight CAT, in 2 transverse rows on rostrum .*cryptocephalus* (Irwin-Smith, 1918) n. comb. (p. 126)

Dracogalerus afrikaanus n. sp.

(Figs. 193 to 194, 196 to 197)

Measurements (1 ♂): L = 1.1 mm; a = 18.3; b = 11.0; c = 10.0; CAT = 17 μm; No CAT = 8; SS = 6-17 μm; first SlAT = 24 μm; last SlAT = 25 μm; first SvAT = 21 μm; last SvAT = 15 μm; No SlAT = 14; No SvAT = 9; Non-ann Term to Tail Length = 45%; T/ABD = 2.7; Spic = 81 μm; Gub = 15 μm.

(Holotype ♀): L = 1.1 mm; b = 7.1; c = 11.4; V = 44%; CAT = 19 μm; No CAT = 8; SS = 7-11 μm; first SlAT = 25 μm; last SlAT = 24 μm; first SvAT = 23 μm; last SvAT = 22 μm; No SlAT = 19; No SvAT = 10; Non-ann Term to Tail Length = 65%; T/ABD = 2.8.

Female Holotype.—Rostrum conoid shaped; base of rostrum expanded, set off from body (fig. 194). Rostrum ornamented with minute punctations; some rostral setae very stout about size of CAT. Amphids loop-shaped (horseshoe-shaped). Eight CAT, 2 sublateral and 2 subdorsal pairs, in 2 transverse rows just anterior to mid-rostrum. Buccal cavity weakly developed. Annules with obscure markings or subcuticular granulation, most prominent on posterior part of body. Two pairs of subventral paravulval setae, 10 to 15 μm long, 1 pair anterior and 1 posterior to vulva. Longest SS on rostrum and mid-body region. Eight rows of SS on esophageal and mid-body regions, 4 sublateral, 2 subdorsal and 2 subventral; 4 sublateral rows on annulated tail region. Without setae intermingled with SlAT. Caudal glands extend anterior to anus 4.0 times ABD. Anterior part of non-annulated tail region with conspicuous punctations, posterior $1/5$ without obvious punctations. Ten pairs of

---

38. *Dracogalerus*, L. m., helmeted dragon.

setae on non-annulated tail region, position measured from last complete tail annule to tail tip; 1 lateral-ventral pair about 75%; 5 subdorsal pairs, 1 pair just posterior to last complete tail annule, 1 pair 1 annule width posterior to first pair, 1 pair just posterior to second pair, 1 pair about 33%, and 1 pair just posterior to fourth pair; 2 latero-dorsal pairs, 1 pair about 33% and 1 about 50%; 2 subventral pairs, 1 pair just posterior to last complete tail annule and 1 about 66%. Tail terminus with long, slender, smooth, spinneret.

Male.—Similar to female. Amphids unispiral. Longest SS on esophageal region. Eight long setae intermingled with SlAT, not alternating with tubes, setae about equal in length to adjacent adhesion tubes. Five unpaired, subventral corniform adhesion tubes posterior to last SvAT 8 to 10 μm long (fig. 196). Caudal glands extend anterior to anus 3.1 times ABD. One pair of stout, broad-based, anal setae, 15 μm long, posterior to anus. Anal opening laterally bounded by oval area of smooth cuticle with obscure punctations, extends about 3 annules. Non-annulated tail region anterior $3/4$ with conspicuous punctations, posterior $1/4$ without punctations. Non-annulated tail region with 3 ventral, rounded protuberants (fig. 196), position measured from last complete tail annule to tail tip; first about 50%, second about 66%, and third about 75%. Six pairs of setae on non-annulated tail region; 1 latero-ventral pair adjacent to second protuberant; 1 lateral to subdorsal pair about 50%; 1 subdorsal pair just posterior to last complete tail annule; 3 subventral pairs, 1 pair just posterior to last complete tail annule, 1 pair adjacent to first protuberant or about 50% and 1 pair located about 1 annule width posterior to second pair.

Holotype (♀).—Collected January 7, 1961 by I. M. Newell. Catalogue No. UCNC 1440, University of California, Davis.

Paratype.—1 ♂. Same data as holotype. Deposited at University of California, Davis (UCNC 1543).

Type Habitat.—Marine, associated with Algae.

Type Locality.—Fifteen miles south of East London, Republic of South Africa.

Distribution.—East London and Hout Bay, Republic of South Africa.

Diagnosis.—Differs from other known species of *Dracogalerus* by conoid shaped rostrum, expanded at base and set off from body.

Dracogalerus bastiani[39] n. sp.

(Figs. 191 to 192, 195)

Measurements (Holotype ♀): L = 1.2 mm; b = 7.6; c = 12.2; V = 44%; CAT = 12 μm; No CAT = 10; SS = 7–19 μm; first SlAT = 25 μm; last SlAT = 22 μm; first SvAT = 21 μm; last SvAT = 20 μm; No SlAT = 28; No SvAT = 15; Non-ann Term to Tail Length = 61%; T/ABD = 2.7.

Female Holotype.—Rostrum broadly rounded anteriorly, base not expanded (fig. 192); rostral setae very stout, about size of CAT; ornamented with vacuoles, most prominent at base, smaller and fewer anteriorly. Amphids partially double spiral, about $1/4$ spiral doubled. Ten CAT, 3 sublateral and 2 subdorsal pairs, in 3 transverse rows on anterior $1/2$ of rostrum. Buccal cavity moderately developed (fig. 192). Annules on mid-body region with obscure markings. Two pairs of subventral paravulval setae, 14 μm long, 1 pair anterior and 1 posterior to vulva. Longest SS on esophageal region. Eight rows of SS on esophageal and mid-body regions, 4 sublateral, 2 subdorsal and 2 subventral; 4 sublateral rows on annulated tail region. Without setae intermingled with SlAT. Caudal glands extend anterior to anus 1.9 times ABD. Anterior $2/3$ of non-annulated tail region ornamented with conspicuous punctations, posterior $1/3$ without punctations; a few small vacuoles just posterior to last complete tail annule. Seven pairs of setae on non-annulated tail region, position measured from last complete tail annule to tail tip; 3 subventral pairs, 1 just posterior to last complete tail annule, 1 about 50%, 1 about 75%; 2 subdorsal pairs, 1 just posterior to last complete tail annule, and 1 about 33%; 2 latero-dorsal pairs, 1 about 75%, and 1 about 66%.

Males.—Not observed.

Holotype (♀).—Collected January 7, 1961 by I. M. Newell. Catalogue No. UCNC 1441, University of California, Davis.

---

39. Named in honor of H. C. Bastian.

Paratypes. —None.
Type Habitat. —Marine, associated with Algae.
Type Locality. —Fifteen miles south of East London, Republic of South Africa.
Distribution. —East London and Hout Bay, Republic of South Africa.
Diagnosis. —Most closely resembles *D. cryptocephalus* n. comb. differs by 10 CAT in 3 transverse rows on rostrum. Differs from *D. afrikaanus* n. sp. by broadly rounded rostrum, not expanded at base; and absence of conspicuously long and slender spinneret.

Dracogalerus cryptocephalus (Irwin-Smith, 1918) n. comb.

Syn: *Notochaetosoma cryptocephala* Irwin-Smith, 1918 n. syn.

Measurements (1 ♂): L = 1.1 mm; a = 16.0; b = ?; c = 7.5; CAT = 15 μm; No CAT = 8; SS = 7–11 μm; SlAT = 26 μm; SvAT = 15 μm; No SlAT = 7; No SvAT = 8; Non-ann Term to Tail Length = 51%; T/ABD = 3.1; Spic = 66 μm; Gub = ?.

Male. —Rostrum broadly rounded anteriorly, base not expanded. Rostrum without ornamentation. Amphids unispiral. Eight CAT, 2 sublateral and 2 subdorsal pairs, in 2 transverse rows about mid-rostrum. Buccal cavity weakly developed, collapsed. Annules without ornamentation. Longest SS on rostrum. Eight rows of SS on esophageal and mid-body regions. Setae present on annulated tail region. Without setae intermingled with SlAT. Subventral corniform adhesion tubes absent. Caudal glands extend anterior to anus about 1.0 times ABD. One subventral pair of anal setae, posterior to anus. Anal opening laterally bounded by oval area of smooth cuticle with small punctations, extends about 2 annules. Anterior $2/3$ of non-annulated tail region ornamented with small punctations, posterior $1/3$ without punctations. Non-annulated tail region with 6 ventral, rounded protuberants, position measured from last complete tail annule to tail tip; first about 20%, second about 30%, third about 40%, fourth about 55%, fifth about 65%, and sixth about 75%. One subventral pair of setae on non-annulated tail region, about 50%, position measured from last complete tail annule to tail tip.

Females. —Not observed.

Type Specimen (1 ♂). —Collected in 1916, by V. A. Irwin-Smith. Australian Museum, Sydney, New South Wales, Australia, No. W458.

Type Habitat. —Marine, collected between high and low tidal zones.

Type Locality. —Vaucluse, Port Jackson, N. S. W., Australia.

Distribution. —Same as type locality.

Diagnosis. —Most closely resembles *D. bastiani* n. sp. differs by 8 CAT in 2 transverse rows on rostrum. Differs from *D. afrikaanus* n. sp. by broadly rounded rostrum, not expanded at base.

Type material of *D. cryptocephalus* was available for study, but these specimens were in poor condition. This species was identifiable from the original description and distinguishable taxonomic characters of the type specimen.

Summary

The superfamily Draconematoidea contains exclusively marine forms. The present revision is based upon the study of over 3,200 specimens obtained from samples collected at 90 locations from the major marine waters in the world. Draconematids appear to be most abundant at depths of less than 50 meters, but some of the species have been recovered at depths of 540 meters in the Antarctica. Most draconematids were found associated with various marine algae. Intraspecific and generic variations of the significant taxonomic characters and the development of the 4 larval stages in 1 genus were studied.

The most profound taxonomic characters of the superfamily are the cephalic and posterior ventral adhesion tubes. These specialized structures are associated with glands, which are presumed to secrete a material that attaches the tubes to the substrate allow-

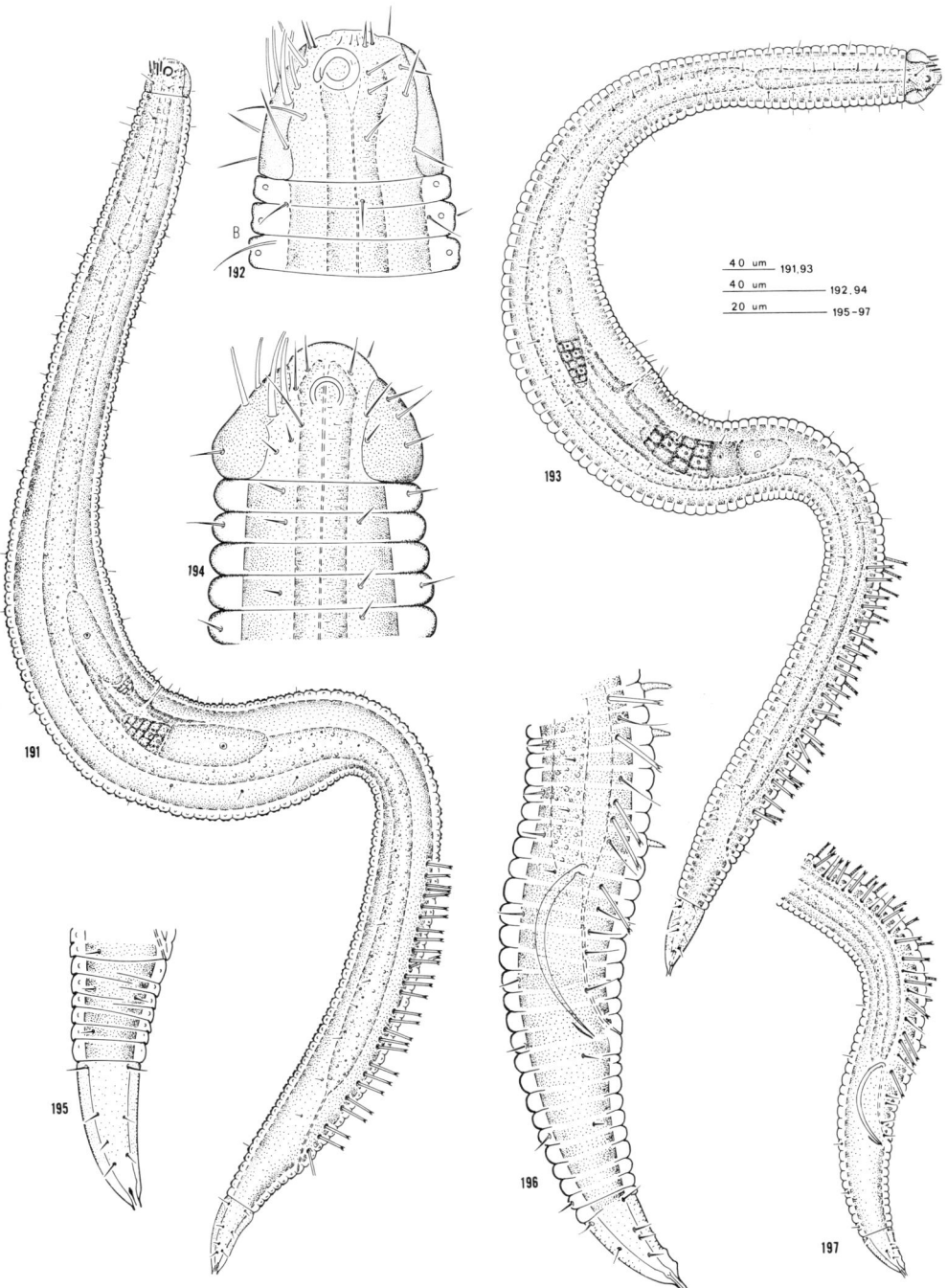

32. Figs. 193–194, 196–197: *Dracogalerus afrikaanus* n. sp. 193) Female, full length; 194) Female, head; 196) Male, tail, note corniform adhesion tubes and ventral protuberants on nonannulated tail region; 197) Male, posterior body region, correct number of corniform adhesion tubes not illustrated. Figs. 191–192, 195: *Dracogalerus bastiani* n. sp. 191) Female, full length, SvAT and correct number of CAT not illustrated; 192) Female, head; 195) Female, tail.

ing the nematode to move with "inchworm-like" movements. Number and size of the tubes vary between species within the different genera and between the immature forms and adults. Development of the adhesion tubes in the various life stages appears to be basically similar in all the species studied, gradually increasing in numbers through the larval stages to adults.

This revision includes 1 new superfamily; 1 new family; 4 new subfamilies; 8 new genera; and 29 new species.

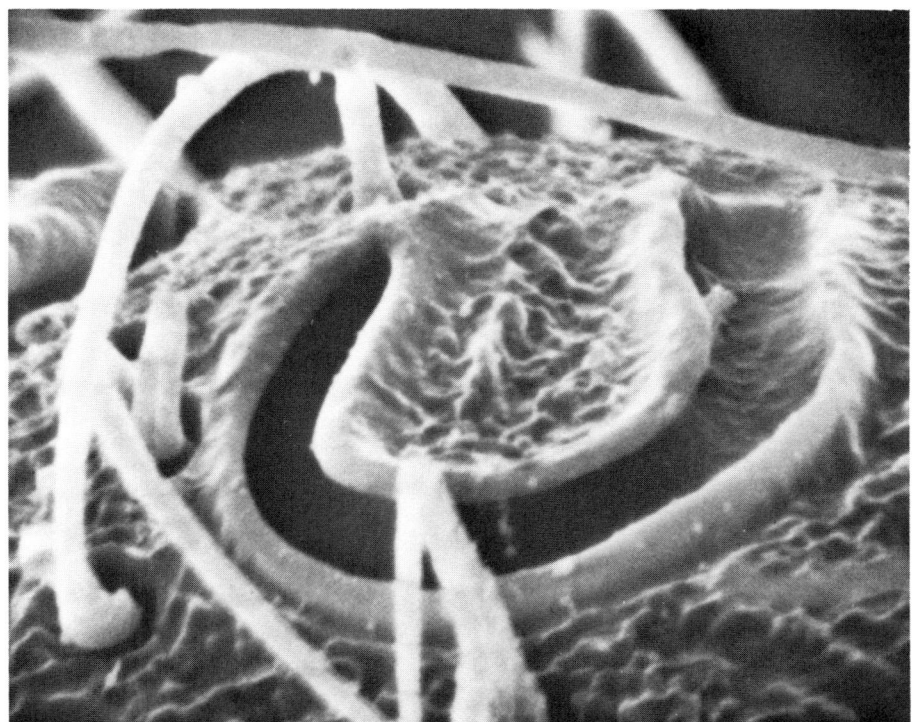

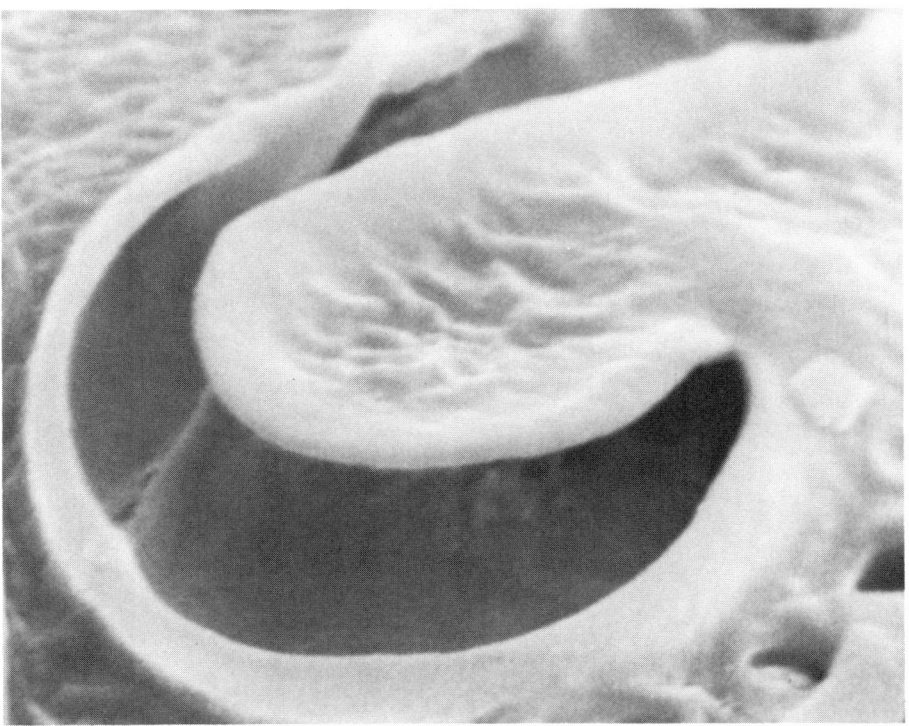

33. FIGS. 198–199: *Draconema cephalatum* Cobb. Scanning electron micrographs of amphidial grooves, note dorsal arm of groove penetrating rostrum appearing as a hole (see Morphology page 11); 198) Male, X 10,000; 199) Female, X 6,500. Prepared by Arnold Bell, Department of Nematology, University of California, Riverside, California.

Literature Cited

ALLGÉN, C.
- 1932. Weitere Beiträge zur Kenntnis der marinen Nematodenfauna der Campbellinsel. Nytt. Mag. Naturvidensk. 70:97-198.
- 1939. Ueber einige im reinen Schalensand der Westküste Norwegens frei lebende Nematoden. Festschr. 60. Geburtst. Embrik Strand 5:404-425.
- 1954. Die Claparedielliden Norwegens. K. Nor. Vidensk. Selsk. Forh. 27(8):37-41.

BAYLIS, H. A., and R. DAUBNEY
- 1926. A synopsis of the families and genera of Nematoda. London. Br. Mus.:1-277.

CHITWOOD, B. G., and M. B. CHITWOOD
- 1937. An introduction to nematology. Sect. I., Pt. I., Baltimore, Md.:1-53.

CLAPARÈDE, J. L. R. A. E.
- 1863. Beobachtungen über Anatomie und Entwicklungsgeschichte wirbelloser Thiere an der Küste von Normandie angestellt. Leipzig:1-120.

COBB, N. A.
- 1913. *Draconema:* A remarkable genus of marine free-living nematodes. J. Wash. Acad. Sci. 3(5):145-149.
- 1929. The ambulatory tubes and other features of the nema *Draconema cephalatum.* J. Wash. Acad. Sci. 19(12):255-260.
- 1933. N. A. Cobb: new nemic genera and species, with taxonomic notes. (Ed. Margaret V. Cobb). J. Parasitol. 20(2):81-94.
- 1935. A key to the genera of free-living nemas. (Eds. Margaret V. Cobb and Corinne Cooper). Proc. Helminthol. Soc. Wash. 2(1):1-40.

CONINCK, L. A. de
- 1965. Systématique des Nématodes. In: Traité de Zoologie, Anatomie, Systématique, Biologie. Nemathelminthes. Masson et Cie, Paris. 4(2):586-681.

DITLEVSEN, H.
- 1915. *Chaetosoma annulatum* n. sp., en repraesentant for gruppen Nematoidea i danskefarvande. Vidensk. Medd. Dan. Naturhist. Foren. Kbh. 66(7):197-203.

FILIPJEV, I. N.
- 1918. Free-living marine nematodes of the vicinity of Sevastopol, Pt. 1. Tr. Osob. Zool. Lab. Sevastopol. Biol. Stantsii Ross. Akad. Nauk. 2(4):1-350.
- 1921. Free-living marine nematodes of the vicinity of Sevastopol, Pt. 2. Tr. Osob. Zool. Lab. Sevastopol. Biol. Stantsii Ross. Akad. Nauk. 2(4):353-614.
- 1934. The classification of the free-living nematodes and their relation to the parasitic nematodes. Smithson. Misc. Collect. (Publ. 3216) 89(6):1-63.

GERLACH, S. A.
- 1952. Nematoden aus dem Küstengrundwasser. Abh. Math.-Naturwiss. Kl., Akad. Wiss. Lit. Mainz 6:315-372.
- 1954. Nouveaux nématodes libres des eaux souterraines littorales francaises. Vie Milieu 4(1):95-110.
- 1957. Die Nematodenfauna des Sandstrandes an der Küste von Mittelbrasilien. (Brasilianische Meeres-Nematoden IV). Mitt. Zool. Mus. Berl. 33(2):411-459.
- 1958. Freilebende Nematoden von den Korallenriffen des Roten Meeres. Kiel. Meeresforsch. 14(2):241-246.

GERLACH, S. A., and F. RIEMANN
- 1973. The Bremerhaven checklist of aquatic nematodes. A catalogue of Nematoda Adenophorea excluding the Dorylaimida. Part I. Veroeff. Inst. Meeresforsch. Bremerhaven Suppl. 4(1):1-404.

GIARD, A., and J. H. BARROIS
- 1874. Note sur un *Chaetosoma* et une Sagitta suivie de quelques reflexions sur la convergence des types par la vie pélagique. Rev. Sc. Nat. Montpellier 3:513-532.

HOPE, W. D., and D. G. MURPHY
  1972. A taxonomic hierarchy and checklist of the genera and higher taxa of marine nematodes. Smithson. Contrib. Zool. 137:1-101.

INGLIS, W. G.
  1968. Interstitial nematodes from St. Vincent's Bay New Caledonia. Expéd. francaise sur les recifs coralliens de la Nouvelle Calédonie, Paris 1967: Editions de la Fondation Singer-Polignac 2:29-74.

IRWIN-SMITH, V. A.
  1918. On the Chaetosomatidae, with descriptions of new species, and a new genus from the coast of New South Wales. Proc. Linn. Soc. N. S. W. 45(4):757-814.

JOHNSTON, T. H.
  1938. A census of the free-living and plant-parasitic nematodes recorded as occurring in Australia. Trans. R. Soc. S. Aust. 62(1):149-167.

KING, L. A. L.
  1939. Marine nematodes. Scott. mar. Biol. Assoc. Annu. Rep. 1937-38:10-13.

KREIS, H. A.
  1928. Weiterer Beitrag zur Kenntnis der freilebenden marinen Nematoden. Arch. Naturg. Berl. 1926, 92 Abt. A. (8):1-29.
  1938. Papers from Dr. Th. Mortensen's Pacific Expedition 1914-16, LXVIII. Neue Nematoden aus der Südsee. Vidensk. Medd. Dan. Naturhist. Foren. Kbh. 1937-38, 101:153-181.
  1963. Marine nematoda. Zool. Iceland 2(14):1-68.

LEVINSEN, G. M. R.
  1882. Smaa Bidrag til den Grønlandske Fauna. Vidensk. Medd. Dan. Naturh. Foren. Kbh. (1881) 4(3):127-136.

LORENZEN, S.
  1973. Die Familie Epsilonematidae (Nematoda). Akad. Wiss. Lit. (Mainz) Math. Naturwiss Kl. Mikrofauna Meeresbodens 25:1-86.

MECHNIKOV, I.
  1867. Beiträge zur Naturgeschichte der Würmer. Z. Wiss. Zool. 17(4):539-544.

MICOLETZKY, H.
  1922. Die freilebenden Erd-Nematoden mit besonderer Berücksichtigung der Steiermark und der Bukowina, zugleich mit einer Revision sämtlicher nicht mariner, freilebender Nematoden in Form von Genus-Beschreibungen und Bestimmungsschlüsseln. Arch. Naturg. Berl. (1921) 87, Abt. A, (8-9):1-650.

MURPHY, D. G.
  1963. Three new species of marine nematodes from the Pacific near Depot Bay, Oregon. Proc. Helminthol. Soc. Wash. 30(2):249-256.

PANCERI, P.
  1876. Osservazioni intorno a nuove forme di vermi nematodi marini (Preliminary note presented 4 Nov.). Rend, Accad. Sci. Fis. Mat. Napoli 15(11):225.
  1878. Osservazioni intorno a nuove forme di vermi nematodi marini. Atti. R. Accad. Sci. Fis. Mat., Napoli 7(10):1-10.

RIEMANN, F.
  1972. Corpus gelatum und ciliäre Strukturen als lichtmikroskopisch sichtbare Bauelemente des Seitenorgans freilebender Nematoden. Z. Morphol. Tiere 72(1):46-76.

SCHEPOTIEFF, A.
  1907. Zur Systematik der Nematoideen. Zool. Anz. 31(5-6):132-161.
  1908. Die Chaetosomatiden. (Untersuchungen über einige wenig bekannte freilebende Nematoden. III). Zool. Jb. (Syst.) 26:401-414.

SCHUURMANS STEKHOVEN, J. H., Jr.
  1935. Nematoda: Systematischer Teil.. Nematoda errantia. In: Grimpe, G., and E. Wagler, Die Tierwelt Nord-Ostsee (Leipzig 1935):1-173.

SHER, S. A., and A. H. BELL
  1975. Scanning electron micrographs of the anterior region of some species of Tylenchoidea (Tylenchida: Nematoda). J. Nematol. 7(1):69-83.

SHIPLEY, A. E.
- 1896. Nemathelminthes. In: Harmer, S. F., and A. E. Shipley (Ed.), The Cambridge Natural History 2:121-194.

SOUTHERN, R.
- 1914. Nemathelmia, Kinorhyncha, and Chaetognatha. (Clare Island Surv. Pt. 54). Proc. R. Ir. Acad. 31(3):1-80.

STEINER, G.
- 1916. Freilebende, Nematoden aus der Barentssee. Zool. Jahrb. Abt. Syst. Oekol. Geogr. Tiere 39(5-6):511-676.
- 1921. Ostasiatische marine Nematoden. Zool. Jahrb. Abt. Syst. Oekol. Geogr. Tiere 44(3):195-226.
- 1927. A new nemic family Epsilonematidae. J. Parasitol. 14(1):65-66.

WITHDRAWN